Transport and Fate of Polycyclic Aromatic Hydrocarbons in Soil-Plant System

Yanzheng Gao
Juan Liu   Fuxing Kang

Science Press
Beijing, China

Responsible Editors: Dan Zhou, Wenhang Liu

Copyright © 2015 by Science Press
Published by Science Press
16 Donghuangchenggen North Street
Beijing 100717, P. R. China

Printed in Beijing

All rights reserved. No part of this publication may be reproduced, stored in a retrieval system, or transmitted in any form or by any means, electronic, mechanical, photocopying, recording or otherwise, without the prior written permission of the copyright owner.

ISBN 978-7-03-045955-8

Preface

Anthropogenic soil contamination has become a worldwide environmental problem during the past decades. Natural and xenobiotic organic pollutants present in soils could be taken up by plants, which is a major pathway for toxic substances to enter food chain. Polycyclic aromatic hydrocarbons (PAHs) are a well-recognized group of organic contaminants that have received vast attention largely due to the concerns on their high toxicity and recalcitrance in soil environment. Hydrophobic characteristics and persistence of PAHs in the environment result in their accumulation and enrichment in soils. PAHs present in soils can be absorbed by plants. Given that plants serve as the foundation of human and animal food webs, daily consumption of PAH-contaminated food could potentially increase human and animal exposure to hazardous substances. Better understanding of transport and fate of PAHs in soil-plant system is therefore essential to protecting human and ecological health from exposure to contaminants.

In the past several years, the research theme of Research Group of Organic Contaminant Control and Soil Remediation, Nanjing Agricultural University has been focusing on the transport and fate of PAHs in soil-plant continuum. The research program received financial supports from National Science Foundation of China (41171193, 41171380, 21477056, 21077056, 40971137, 40701073, 20777036, 20507009), Science Foundation of Jiangsu Province (BK20130030, BK2009315, BK2007580, BK2006518), and Special Fund for Agro-scientific Research in Public Interest, China (201503107). The major findings are incorporated in this book.

The book composed of two parts, Part I Soil and Part II Plant. Part I focuses on the fractions, availability and rhizospheric distributions of PAHs in soil environments. We elucidated the partitioning process of PAHs among soil, water and plant roots. During plant growth, roots could actively or passively release a range of organic compounds referred to as root exudates. Low-molecular-weight organic acids (LMWOAs) occurring widely in soils are a group of natural products present in root exudates. In Part I, the impacts of root exudates and LMWOAs to sorption/desorption, release and availability of PAHs in soils were elucidated. Part II focuses on uptake, subcellular distributions, and metabolism of PAHs in plants. Microorganisms associated with plants play a key role in PAH uptake by plants. Plant-arbuscular mycorrhizal fungi (AMF) symbioses are ubiquitous in the environment. We investigated the influence of AMF on plant uptake and accumulation of PAHs from soils, and evaluated the functions of AMF hyphae in the PAH uptake by plants. Endophytic bacteria form one of the microbial communities most closely associated with plants. Colonization of PAH-degrading endophytic bacteria provided a novel method for removal of PAHs within plants.

I gratefully thank Wanting Ling, Xuezhu Zhu, and Huoliang Kong for their participation in writing this book. Thanks are also given to Kai Sun, Anping Peng, Binqing Sun, Yizeng Wang, Nan Wang, Rui Sun, Hongjiao Dang, Lili Ren, Dongsheng Chen, Yi Zhang, Shuaishuai Gong, Yan Yang, Xiaodan Lu, Yuechun Zeng, Zhaoxia Cheng, Qiuling Li, Wei Xiong, and Xiaojia Yuan for their research contributions.

Yanzheng Gao

October 2015

Contents

Preface

PART I SOIL

Chapter 1 The forms and availability of PAHs in soil ·· 3
  1.1 The forms of PAHs in soil ··· 3
      1.1.1 Fractionation methods of PAH residues in soil ································· 4
      1.1.2 Desorbing fraction of PAHs in soil ·· 5
      1.1.3 Non-desorbing fraction of PAHs in soil ·· 7
      1.1.4 Bound residues of PAHs in soil ·· 10
  1.2 The availability of PAHs in soil ·· 12
      1.2.1 Available fractions of PAHs in soils as a function of aging time ············ 13
      1.2.2 Microbial degradation of available fractions of PAHs in soils ··············· 15
      1.2.3 Transformation of available fraction of PAHs to bound residue in soils ····· 17
      1.2.4 Butanol-extraction technique for predicting the availability of PAHs in soil ····· 18
      1.2.5 Phytoavailability of bound-PAH residues in soils ····························· 18

Chapter 2 Gradient distribution of PAHs in rhizosphere soil ······························ 24
  2.1 Gradient distribution of PAHs in rhizosphere soil: a greenhouse experiment ·· 25
      2.1.1 Gradient distribution of phenanthrene and pyrene in rhizosphere ············ 26
      2.1.2 Gradient distribution of root exudates in rhizosphere ························ 27
      2.1.3 The correlations of PAH concentration gradient with the concentration gradient of root exudates in rhizosphere ······································· 30
  2.2 *In situ* gradient distribution of PAHs in rhizosphere soil: a field study ······· 33
      2.2.1 *In situ* gradient distribution of PAHs in rhizosphere soil ···················· 34
      2.2.2 Rhizosphere effects on PAH distribution in soil ······························ 37
  2.3 Rhizospheric gradient distribution of bound-PAH residues in soils ············ 40
      2.3.1 Gradient distribution of bound-PAH residues in rhizosphere ················ 41
      2.3.2 Mechanism of rhizospheric gradient distribution of bound-PAH residues in soils ···· 46

Chapter 3 Partition of PAHs among soil, water and plant root ·· 48
   3.1   Sorption of PAHs by soils with heavy metal co-contaminants ················ 49
       3.1.1   Sorption isotherms of phenanthrene by soils ···································· 50
       3.1.2   Sorption of phenanthrene by heavy metal-contaminated soils ·············· 52
       3.1.3   Mechanisms of the heavy metal enhanced-sorption of phenanthrene by soils ···· 53
   3.2   Dissolved organic matter (DOM) influences the partition of PAHs between soil and water ··· 57
       3.2.1   Effect of inherent DOM on phenanthrene sorption by soils ················ 59
       3.2.2   Effect of exotic DOM on phenanthrene sorption by soils ···················· 63
   3.3   Partition of polycyclic aromatic hydrocarbons between plant root and water ··· 67
       3.3.1   Partition of phenanthrene between roots and water ·························· 68
       3.3.2   Estimation of partition coefficient of phenanthrene between root and water using a composition model ·· 70
       3.3.3   Partition of phenanthrene between root cell walls and water ·············· 71

Chapter 4 Impact of root exudates on the sorption, desorption and availability of PAHs in soil ·· 73
   4.1   Impact of PAHs on root exudate release in rhizosphere ······················ 73
       4.1.1   Impact of PAH contamination levels on root exudation in rhizosphere ············ 74
       4.1.2   Distribution of root exudates in different layers of rhizosphere soil ········ 77
   4.2   Impact of root exudates on PAH sorption by soils ······························· 77
       4.2.1   Root exudate component-influenced sorption of PAH by soil ············ 78
       4.2.2   Mechanism discussions ·· 80
   4.3   Impact of root exudates on PAH desorption from soils ······················· 83
       4.3.1   Desorption of PAHs from soils as a function of root exudate concentration ············ 84
       4.3.2   PAH desorption by root exudates in different soils ···························· 86
       4.3.3   Effects of soil aging on PAH desorption by root exudates from soil ······ 87
       4.3.4   Desorption of different PAHs by root exudates in soil ························ 88
       4.3.5   Impact of root exudate components on PAH desorption in soil ··········· 89
       4.3.6   Dissolved organic matter in soils with the addition of root exudates ······ 90
   4.4   Impact of root exudates on PAH availabilities in soils ························ 92
       4.4.1   Impact of root exudates on n-butanol-extractable pyrene in soil ··········· 93
       4.4.2   Impact of root exudate components on the n-butanol-extractable pyrene in soil ·· 95
       4.4.3   Mechanisms by which root exudate and its components influence PAH availa-

bility in soil ··· 98

Chapter 5 Low-molecular-weight organic acids (LMWOAs) influence the transport and fate of PAHs in soil ·· **101**

5.1 LMWOAs-influence the PAH sorption by different soil particle size fractions ··· 102

  5.1.1 Fractionation protocol of different soil particle size fractions ············· 103
  5.1.2 PAH sorption by different soil particle size fractions ·························· 106
  5.1.3 Effects of LMWOAs on PAH sorption by different soil particle size fractions ··· 108
  5.1.4 Mechanisms of LMWOA-influenced PAH sorption by different soil particle size fractions ·· 109

5.2 LMWOAs enhance the PAH desorption from soil ·······························114

  5.2.1 LMWOA-enhanced desorption of PAH from PAH-spiked soil ············115
  5.2.2 LMWOA-enhanced desorption of PAHs from soils collected from a PAH-contaminated site ··118
  5.2.3 Mechanisms of LMWOA-enhanced desorption of PAHs from soils ····124

5.3 Impact of LMWOAs on the availability of PAHs in soil ···················· 127

  5.3.1 Impact of LMWOAs on the butanol-extractable PAHs in soils ············128
  5.3.2 Mechanism discussions ···132

5.4 Elution of soil PAHs using LMWOAs ·· 133

  5.4.1 Elution of PAHs in soil columns by LMWOAs ····································135
  5.4.2 Distributions of PAHs in soil columns ···136
  5.4.3 Butanol-extractable and nonextractable PAHs in soil columns ···········137
  5.4.4 Impact of soil type on PAH elution ··138
  5.4.5 Mechanisms of LMWOA-enhanced elution of soil PAHs ····················140
  5.4.6 Relationship between the elution of PAHs and the dissolution of metal ions ··· 141

5.5 LMWOAs enhance the release of bound PAH residues in soil ················ 146

  5.5.1 The release of bound PAH residues in soils as a function of incubation time···147
  5.5.2 LMWOA-enhanced release of bound PAH residues in soil ··················149

PART II PLANT

Chapter 6 Uptake, accumulation and translocation of PAHs in plants ············ **157**

6.1 Uptake pathways of PAHs in plants ·· 158

  6.1.1 Root uptake of PAHs ··160

6.1.2　Shoot accumulation of PAHs ··161
　　　6.1.3　Uptake time and PAH concentration influence their uptake by plants···········164
　6.2　Accumulation and translocation of PAHs in plants with different
　　　compositions ··· 166
　　　6.2.1　Accumulation of PAHs in roots ···167
　　　6.2.2　Accumulation of PAHs in shoots ···171
　　　6.2.3　Translocation of PAHs in plant ···173
　6.3　Comparison for plant uptake of PAHs from soil and water ····················· 174
　　　6.3.1　Plant uptake of PAHs from water···175
　　　6.3.2　Plant uptake of PAHs from soil···177
　　　6.3.3　Comparison for plant uptake of PAHs from soil and water····················178

Chapter 7　Subcellular distribution of PAHs in plants ·······························181
　7.1　PAH distribution in subcellular root tissues ·· 181
　　　7.1.1　Fractionation protocol of root subcellular tissues ··································182
　　　7.1.2　Uptake of PAHs by roots ···182
　　　7.1.3　Subcellular movement and distribution of PAHs in root cells·················185
　7.2　Subcellular distribution of PAHs in arbuscular mycorrhizal roots ············ 188
　　　7.2.1　PAH concentrations in subcellular tissues of arbuscular mycorrhizal roots······189
　　　7.2.2　Subcellular concentration factors of PAH in arbuscular mycorrhizal roots ······191
　　　7.2.3　Proportion of PAH in subcellular tissues of arbuscular mycorrhizal roots ······192

Chapter 8　Metabolism of PAHs in plants···194
　8.1　Metabolism of anthracene in tall fescue·· 194
　　　8.1.1　Metabolism of anthracene in tall fescue ···195
　　　8.1.2　Distribution of anthracene and its metabolites in subcellular tissues················198
　　　8.1.3　Metabolism mechanism discussion ··204
　8.2　Enzyme activity in tall fescue contaminated by PAHs ···························· 205
　　　8.2.1　Enzyme activity in tall fescue ··206
　　　8.2.2　Enzyme activity in subcellular fractions of tall fescue···························207
　8.3　Inhibitor reduces enzyme activity and enhances PAH accumulation in tall
　　　fescue···211
　　　8.3.1　*In vitro* degradation of PAHs in solution with enzymes ························213
　　　8.3.2　Effects of inhibitor on enzyme activities in plants ································215
　　　8.3.3　Effects of inhibitor on the enhanced accumulation of PAH in plants ·····217

Chapter 9　Arbuscular mycorrhizal fungi influence PAH uptake by plants ·····221
　9.1　PAH uptake by arbuscular mycorrhizal plants ·· 221

   9.1.1 Arbuscular mycorrhizal colonization of root exposed to PAHs in soil ············222

   9.1.2 PAH uptake by arbuscular mycorrhizal plants ································223

 9.2 Arbuscular mycorrhizal hyphae contribute to PAH uptake by plant ········ 226

   9.2.1 Three-compartment systems ··227

   9.2.2 Mycorrhizal root colonization and plant biomass ························228

   9.2.3 Concentrations of PAHs in mycorrhizal roots ····························229

   9.2.4 Partition coefficients of PAHs by arbuscular mycorrhizal hyphae ···············230

   9.2.5 Translocation of PAHs by arbuscular mycorrhizal hyphae ·············233

Chapter 10 Utilizing PAH-degrading endophytic bacteria to reduce the plant PAH contamination ··236

 10.1 Distribution of endophytic bacteria in plants from PAH-contaminated soils ·· 237

   10.1.1 PAH concentrations in plants from PAH-contaminated soils ············238

   10.1.2 Endophytic bacterial community in PAH-contaminated plants ·······241

   10.1.3 Cultivable endophytic bacterial populations in PAH-contaminated plants ····245

   10.1.4 Amounts of cultivable endophytic bacteria in PAH-contaminated plants ·······246

 10.2 Inoculating plants with the endophytic bacterium *Pseudomonas* sp. Ph6-*gfp* to reduce phenanthrene contamination ···································· 249

   10.2.1 Isolation, identification, and *gfp*-labeling of *Pseudomonas* sp. Ph6 ··············250

   10.2.2 Biodegradation of phenanthrene by Ph6-*gfp* in culture solution ·············252

   10.2.3 Colonization and distribution of Ph6-*gfp* in plants ·····················253

   10.2.4 Performances of Ph6-*gfp* mediate the uptake of phenanthrene by plants ·······258

 10.3 Utilizing endophytic bacterium *Staphylococcus* sp. BJ06 to reduce plant pyrene contamination ··· 262

   10.3.1 Isolation and identification of *Staphylococcus* sp. BJ06 ··············262

   10.3.2 Biodegradation of pyrene by BJ06 in culture solution ················263

   10.3.3 Reducing plant pyrene contamination using strain BJ06 ··············266

References ···**270**

Plates

PART I SOIL

Chapter 1   The forms and availability of PAHs in soil

Soil is considered to be one of the most important natural resources for human beings. However, organic pollutants occur frequently within the soil environment as a result of air deposition, sewage irrigation, and industrial accidents (Gao et al., 2009). This organic pollution triggered by human activities has been a long-term environmental problem in past decades (Führ and Mittelstaedt, 1980; Kipopoulou et al., 1999; Gao et al., 2007). Because of the health hazards of these organic contaminants, knowledge on their transport and fate in the soil environment is of crucial importance in dealing with contaminated sites.

As priority pollutants that are commonly found in the soil environment, polycyclic aromatic hydrocarbons (PAHs) are of major concern due to their recalcitrance and strong mutagenic/carcinogenic properties (Weber and Huang, 2003; Tang et al., 2007). The hydrophobic characteristic and persistence of PAHs result in their accumulation and enrichment in soils. PAHs are widespread and occur at high concentrations of hundreds of mg/kg in soils of many countries (Joner and Leyval, 2003; Ling et al., 2013). Contamination of soil with PAHs poses risks to human and ecosystem health.

When entering into soils, a significant proportion of the organic contaminants is not extractable, but is found bound to soil solids. These bound contaminant residues are less available for plant uptake (Ling et al., 2010). Researchers now realize that data on only the extractable or total concentrations of a given organic chemical may be of limited utility when assessing its environmental significance (Macleod and Semple, 2003). Instead, the form and availability of these contaminants in soil are the most important indices for risk assessment.

1.1   The forms of PAHs in soil

The forms of organic contaminants in soil environments have been reported in literatures (Monteiro et al., 1999; Northcott and Jones, 2000; Loiseau and Barriuso, 2002; Lesan and Bhandari, 2004). Macleod and Semple (2003) observed that the extractable fraction of pyrene decreased significantly, whereas the bound residue increased with its contact time in soil. Similar results were observed by other researchers (Kohl and Rice, 1998; Käcker et al., 2002). However, the PAH concentrations tested in these studies were at their native concentrations in soils, which may be far lower than those at contaminated sites. In addition, only a very limited number of PAHs and soils have been investigated thus far, while the interactions between the forms of PAHs and the influences of soil properties as well as other environmental factors, such as microbial activity on PAH forms still remain unclear. Recently, we fractionated the forms of parent PAH compounds in soils (Gao et al., 2009; Ling et al., 2010). The influence of aging time and microbial activities on the forms of PAHs was also investigated. Results of this work will have considerable benefits for risk assessment, food security, and development of remediation strategies for contaminated sites.

1.1.1 Fractionation methods of PAH residues in soil

A sequential extraction/chemical mass balance approach described by Sabaté et al (2006) was used to fractionate the forms of parent PAH compounds in soils. PAHs in soil were separated into three fractions: a desorbing fraction, a non-desorbing fraction, and a bound residual fraction (Gao et al., 2009; Ling et al., 2010).

(1) Desorbing fraction. A mild extraction technique to obtain the desorbing fraction of PAHs was adapted according to the methods described by Reid et al. (2000) and Cuypers et al. (2002). Three grams of treated soil from each microcosm were placed in a 25 mL glass centrifuge tube, and 15 mL of the mild extraction solution were added. Mild extraction solution consisted of 70 mmol/L hydroxypropyl-β-cyclodextrin (HPCD) and 0.05 g NaN_3 per mL in Milli-Q water. Tubes were closed with a Teflon-liner cap, shielded from light, and shaken horizontally at 150 r/min at 25℃. At 60 h, 120 h and 240 h, tubes were centrifuged for 25 min at 2000 r/min to separate soil from aqueous solution. The supernatant was collected, and fresh mild extraction solution was added. Tubes were then shaken and centrifuged again. The supernatant was liquid-liquid extracted three times using 10 mL of dichloromethane, and the extraction efficiency was tested. Organic phases were dehydrated by percolation through Na_2SO_4 anhydride and combined. The solvent was firstly concentrated by rotary evaporation, then evaporated under a gentle stream of N_2, and diluted with methanol to a final volume of 2 mL. After filtration through a 0.22 μm filter, PAHs were detected by high pressure liquid chromatography (HPLC).

(2) Non-desorbing fraction. This fraction was obtained by exhaustive extraction

following mild extraction. After 240 h of mild extraction for the desorbing fraction, the pellet (soil) was dried at 37℃ for 24 h. Dried soil was then placed in a 25 mL glass centrifuge tube, and 10 mL of a solution of dichloromethane : acetone (1 : 1 vol/vol) were added. Extractions were conducted four times in an ultrasonic bath for 10 min. Soil and solvent were separated by centrifugation for 25 min at 2000 r/min, and then treated as described above.

(3) Bound residue extraction. Dried soil samples resulting from exhaustive extractions were extracted in order to obtain the bound residue fraction. The extraction method was as described by Richnow et al. (2000). After exhaustive extraction, soil samples were placed in glass vials. A 10 mL solution of 2 mol/L NaOH was added to each vial. The vials were closed with Teflon-lined caps and then heat-treated at 100℃ for 2 h. The aqueous fraction was obtained by centrifugation at 2000 r/min for 25 min, acidified with 6 mol/L HCl to a pH < 2, and liquid-liquid extracted three times with 10 mL of dichloromethane. The samples were then treated as described above.

1.1.2 Desorbing fraction of PAHs in soil

The desorbing fraction of a compound in soils has been shown to be the most bioavailable portion (Reid et al., 2000; Cuypers et al., 2002). In this study, we utilized water in combination with HPCD to extract the desorbing fractions of PAHs in a yellow-brown soil collected from Nanjing, Jiangsu. The soil type is a Typic Paleudalf, a typical zonal soil in East China, with a pH of 6.02, 14.3 g/kg soil organic carbon content (f_{oc}), 24.7% clay, 13.4% sand, and 61.9% silt. Test PAHs included fluorene (FLU), phenanthrene (PHE), fluoranthene (FLT), pyrene (PYR), benzo[a]anthracene (BaA), and benzo[a]pyrene (BaP). Cyclodextrins have a hydrophilic cavity but also contain a hydrophobic organic cavity within their molecular structure, which allows the formation of a water-soluble inclusion complex between the cyclodextrin molecule and low polarity organic compounds (Shieh and Hedges, 1996). Successful use of HPCD for estimation of the desorbing PAH fraction and correlation with the biodegradable fraction of organic compounds in a few soils and sediments have been reported (Reid et al., 2000; Cuypers et al., 2002).

The concentrations of the desorbing fraction of PAHs in soil over 0~16 week were given in Figure 1-1. As shown, the concentrations of the desorbing fraction of PAHs clearly decreased after 16 weeks, and were only 11.8%~67.0% of their initial concentrations of this fraction in non-sterilized soil. However, the decrease magnitude of this fraction varied greatly for different PAHs, and 85.3%, 88.2%, 78.8%, 69.1%,

33.0%, and 41.0% of the desorbing fractions of FLU, FHE, FLT, PYR, BaA, and BaP dissipated after 16 weeks. Clearly, this order was inversely correlated with the molecular weights and benzene-ring numbers of the tested PAHs. Concentrations of the desorbing PAHs fraction were much higher at 16 weeks in sterilized soils versus nonsterilized treatments. On a whole, the concentrations of this fraction in sterilized soils decreased to some extend after 4 weeks, and then remained nearly constant in 8~16 week-incubation (Figure 1-1). The former decrease after 4 weeks in this fraction in sterilized soils may be attributed to transfer of "easily desorbing sites" to "difficultly desorbing sites" and "irreversible sites" (Sun and Li, 2005). Since the desorbing fractions of organic compounds are the most bioavailable, the above results indicate that synergistic microbial degradation dominated the dissipation of desorbing fractions of PAHs in the soil environment.

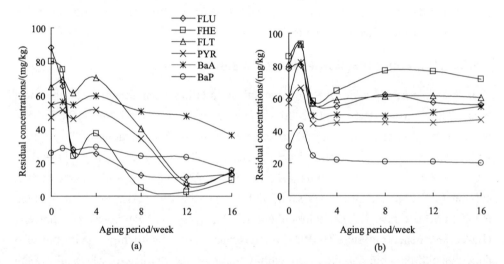

Figure 1-1 Concentrations of the desorbing fraction of PAHs in non-sterilized (a) and sterilized (b) soil as a function of time

Interestingly, as tabulated in Table 1-1, the dissipation amount of the desorbing fractions of the tested PAHs was about 76.1%~152% of their total dissipation in soils (non-sterilized soil as an example). This means that the desorbing fraction was most easily degraded, and that degradation of this fraction contributed predominantly to the total dissipation of PAHs in soils. However, not all of the decrease in the desorbing fractions of PAHs was ascribed to microbial biodegradation. In fact, some of this fraction could transfer to other fractions (such as non-desorbing and bound residual

Table 1-1 The concentrations of the total and desorbing fraction of PAHs in soils

PAHs	$C_{0\text{-total}}$ /(mg/kg)	$C_{16\text{weeks-total}}$ /(mg/kg)	ΔC_{total} /(mg/kg)	$C_{0\text{-HPCD}}$ /(mg/kg)	$C_{16\text{weeks-HPCD}}$ /(mg/kg)	$\Delta C_{\text{total-HPCD}}$ /(mg/kg)	$\Delta C_{\text{total-HPCD}}/\Delta C_{\text{total}}$ /%
Fluorene	93.82	15.09	78.73	88.28	13.01	75.27	95.61
Phenanthrene	87.78	10.91	76.87	80.08	9.42	70.67	91.93
Fluoranthrene	82.12	20.82	61.30	65.06	13.80	51.26	83.63
Pyrene	66.34	23.77	42.57	46.85	14.47	32.38	76.07
Benzo[a]anthracene	77.17	65.40	11.78	54.10	36.24	17.86	151.7
Benzo[a]pyrene	52.50	42.71	9.79	25.64	15.13	10.51	107.4

Note: $C_{0\text{-total}}$ and $C_{16\text{weeks-total}}$ were the concentrations of the total PAH contents at 0 week and 16 weeks, respectively; $C_{0\text{-HPCD}}$ and $C_{16\text{weeks-HPCD}}$ were the concentrations of the desorbing fraction of PAHs in soils at 0 week and 16 weeks, respectively. ΔC_{total} and $\Delta C_{\text{total-HPCD}}$ were the dissipation of the total and desorbing fraction of PAHs at 0 week and 16 weeks, respectively; $\Delta C_{\text{total}} = C_{0\text{-total}} - C_{16\text{weeks-total}}$; $\Delta C_{\text{total-HPCD}} = C_{0\text{-HPCD}} - C_{16\text{weeks-HPCD}}$.

fractions). As seen in Table 1-1, the dissipation amount of the desorbing fraction of BaA and BaP was more than 100% (107.4% and 151.7%, respectively) of their total dissipation in soils. Thus, it is highly likely that parts of their desorbing fractions transferred to other forms in the soil environment.

The percentage of the desorbing fraction relative to the total contents of PAHs at specific time points was calculated and illustrated in Figure 1-2. As seen, this percentage decreased from 94.1%, 91.2%, 79.2%, 70.6%, 70.1%, and 48.8% to 86.2%, 86.2%, 66.3%, 60.9%, 55.4%, and 35.4% after 16 weeks of aging in non-sterilized for FLU, FHE, FLT, PYR, BaA, and BaP, respectively. However, this percentage was slightly higher in the sterilized control soils. As stated, the decrease in this percentage over the 0~16 week can also be ascribed to both microbial degradation of this fraction and its transfer into other fractions in the soil.

1.1.3 Non-desorbing fraction of PAHs in soil

The non-desorbing fractions of the six PAHs in soil as a function of time are given in Figure 1-3. Concentrations of this fraction of PAHs with lower molecular weight generally decreased in non-sterilized soils over 0~16 week. For instance, the concentrations of the non-desorbing fractions of FLU, FHE, FLT, and PYR after a 16 weeks aging decreased from 4.74 mg/kg, 7.53 mg/kg, 16.6 mg/kg, and 19.0mg/kg to 2.06 mg/kg, 1.49 mg/kg, 6.74 mg/kg, and 9.19mg/kg, respectively. However, for

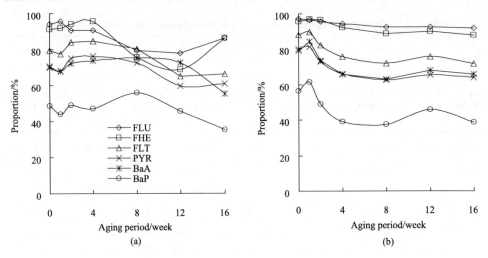

Figure 1-2 The proportion of the desorbing fraction to the total contents of PAHs in non-sterilized (a) and sterilized (b) soil as a function of time

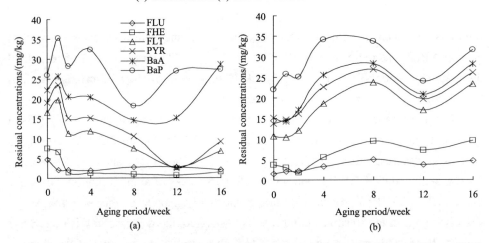

Figure 1-3 Concentrations of the non-desorbing fraction of PAHs in non-sterilized (a) and sterilized (b) soil as a function of time

BaA and BaP with 5 benzene rings and high molecular weights, the concentrations of this fraction were almost constant over the time period, reflecting the recalcitrant nature of these compounds.

In sterilized control soils, the concentrations of the non-desorbing fraction of PAHs after 16 weeks increased to 4.73 mg/kg, 9.62 mg/kg, 23.4 mg/kg, 26.1 mg/kg, 28.2 mg/kg, and 31.7 mg/kg for FLU, FHE, FLT, PYR, BaA, and BaP, respectively. These changes were the opposite of those observed for the non-sterilized soils

(Figure 1-3(a)), indicating the major contribution of the microbial degradation to the dissipation of this fraction of PAHs in non-sterilized treatments. In addition, microbial activities were excluded in sterilized soils, and biodegradation of the desorbing fractions of PAHs could be negligible, while the desorbing fractions of PAHs may partially transfer into other fractions such as the non-desorbing fraction, resulting in an increase in the concentrations of non-desorbing fraction in sterilized control treatments.

The percentage of the non-desorbing fractions relative to their total contents at specific time points was calculated (Figure 1-4). This percentage increased over 0~16 week irrespective of sterilized and non-sterilized soils, which was the opposite of the trend for the desorbing fractions of PAHs in these soils (Figure 1-2). For instance, the percentage for FLU, FHE, FLT, PYR, BaA, and BaP in non-sterilized soils (Figure 1-4(a)) after 16 weeks increased from 5.05%, 8.57%, 20.2%, 28.6%, 28.9%, and 49.5% to 13.6%, 13.62%, 32.4%, 38.7%, 43.8%, and 64.0%, respectively. This percentage was higher for all of the tested PAHs in sterilized versus non-sterilized soils. In sterilized treatments, since microbial activities were excluded, the increase in this percentage could be ascribed to the transfer of the desorbing fraction. However, in non-sterilized soils, most of the desorbing PAH fractions were degraded (as described earlier), consequently resulting in the higher proportion of non-desorbing PAH residues after 16 weeks. In addition, this percentage was positively correlated with the molecular weights of the tested PAHs (Figure 1-4(b)), indicating that the larger PAHs were more likely to be present in the non-desorbing fractions.

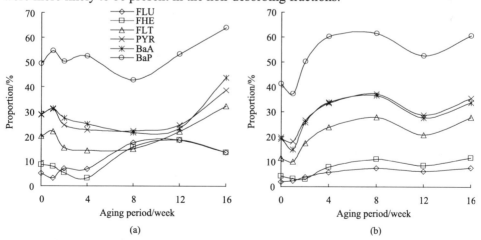

Figure 1-4 The proportion of the non-desorbing fraction to the total contents of PAHs in non-sterilized (a) and sterilized (b) soil as a function of time

One notes that, as displayed in Figure 1-2 and Figure 1-4, the desorbing and non-desorbing fractions always dominated the total residue of PAHs, and more than 96% of the fluorene, 86% of the phenanthrene, 85% of the fluoranthene, 83% of the pyrene, 96% of the benzo[a]anthracene, and 98% of the benzo[a]pyrene existed as extractable fractions (including desorbing and non-desorbing fractions).

1.1.4 Bound residues of PAHs in soil

The concentrations of the bound PAH residue in soil during the 0~16 week incubation were displayed in Figure 1-5. As shown, the concentrations of the PAH-bound residues were much lower than those of the desorbing and non-desorbing fractions (Figures 1-1 and Figure 1-3). The concentrations of bound PAH residues generally tended to increase first and decrease thereafter over 0~16 week. At early stages, the increase in this PAH fractions could result from the transfer of other PAH fractions as described above. The obvious decreases in the desorbing and non-desorbing PAH fractions over the 0~16 week as shown in Figure 1-1(a) and Figure 1-3(a) supported this hypothesis. However, the bound PAH residues tended to decrease after 8~12 week.

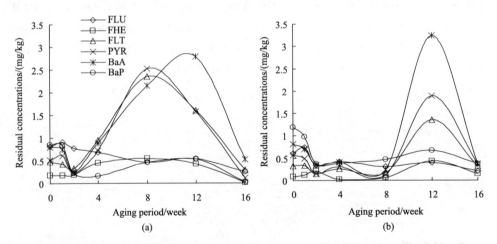

Figure 1-5 Concentrations of the PAH-bound residue in non-sterilized (a) and sterilized (b) soil as a function of time

As reported by other researchers, the bound residues of organic chemicals may be conditionally released into the soil environment (Jerzy et al., 1997; Ying et al., 2005). Here, since the desorbing fractions of PAHs were sharply reduced after 16 weeks primarily due to microbial degradation, the bound PAH residues may be released and become part of the desorbing and non-desorbing PAH fractions. On the other hand, the newly observed PAH-bound residues from the early stages are still unstable (Gevao et al., 2000), and they may again change to other forms of PAHs. As seen in the sterilized control soils (Figure 1-5(b)), the concentrations of the bound PAH residues obviously decreased after 12~16 week after the early stages of increase. However, during this period, the desorbing fraction and the total contents of the PAHs remained nearly constant, while the non-desorbing PAH fractions increased. This indicated that the bound PAH residues may partially transfer into the non-desorbing fractions between 12 weeks and 16 weeks.

The percentage of the bound PAH residues relative to the total contents of PAHs at specific time points was calculated and shown in Figure 1-6. This percentage tended to first increase, and then decrease thereafter from 0 week to 16 weeks irrespective of non-sterilized or sterilized control soils. Compared to the desorbing and non-desorbing PAH fractions, these percentages were much lower, less than 16% and 4.5% for the non-sterilized and sterilized soils, respectively. In addition, this percentage was higher in non-sterilized soils versus sterilized controls. The influence of microbial activities on aging and sequestration of hydrophobic organic chemicals in soil is an area of growing interest (Carmichael et al., 1997; Guthrie and Pfaender, 1998; Gao et al., 2009). In this experiment, biological activity was shown to play an important role in the formation of bound PAH residues. This is similar to findings of previous studies demonstrating that the presence of microbes would lead to the formation of larger solvent non-extractable residues (Macleod and Semple, 2003).

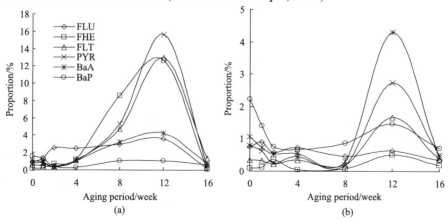

Figure 1-6 The proportion of the bound residue of PAH to its total contents in non-sterilized (a) and sterilized (b) soil as a function of time

In summary, using a sequential extraction mass balance approach, the parent PAH compounds in soil were fractionated into three fractions: a desorbing fraction, a non-desorbing fraction, and a bound residual fraction. More than 83% of each test PAH was extractable, and the desorbing and non-desorbing fractions always dominated the total PAH contents in soils. The portions of bound PAH residues were very small (<16%), and even smaller (4.5%) in sterilized soils. During the aging period of 0~16 week, the desorbing fraction of PAHs in tested soils clearly decreased, and microbial degradation contributed predominantly to the decrease in this fraction. The concentrations of the non-desorbing fraction of PAHs increased in sterilized soils, and remained nearly constant or decreased to some extent in non-sterilized soils. However, the percentage of this fraction relative to the total content of PAHs significantly increased in 0~16 week, and this percentage was correlated with the molecular weights of the test PAHs, indicating that PAHs with larger molecular weights were more likely to be present in the non-desorbing fraction. The bound PAH residues tended to increase at first, and decrease thereafter over the 0~16 week, and microbes contributed to the formation of bound residues in soils.

1.2 The availability of PAHs in soil

HOCs in soil can be fractionated into two groups: the extractable fraction and bound residue. The former, including desorbing and non-desorbing fractions, is the available fraction of a compound in soil that can be absorbed by plants and/or soil-inhabiting animals, while the latter is generally much less available for plant uptake (Führ and Mittelstaedt, 1980; Jerzy et al., 1997). The available concentration of an organic pollutant is less than its total concentration in soil (Noordkamp et al., 1997; Tang et al., 2002; Ling et al., 2010). Therefore, it is inappropriate to use the total concentration as a measure of potential exposure of plants or soil organisms. Instead, availability of the contaminant is a better measure of its potential exposure (Alexander 2000; Semple et al., 2003).

This study was performed to characterize the temporal changes in extractability and to evaluate the availabilities of PAHs in several aging soils. The general properties of test four soils were given in Table 1-2. The results of this study have important applications in risk assessment of PAH-contaminated soils, as well as in the development of remedial strategies for contaminated sites.

Table 1-2 some properties of test soils

Soil No.	Location	Soil texture	pH value	f_{oc} /(g/kg)	Clay content /%	Sand content /%	Silt content /%
Soil 1	Jiangxi	Sandy clay loam	4.56	9.97	36.8	40.7	22.5
Soil 2	Hubei	Silt clay loam	4.74	9.47	39.2	9.20	51.6
Soil 3	Jiangsu	Silt loam	6.02	14.3	24.7	13.4	61.9
Soil 4	Shandong	Loam	7.35	6.92	18.4	34.7	46.9

Note: f_{oc} - Soil organic carbon content.

1.2.1 Available fractions of PAHs in soils as a function of aging time

The available fractions including desorbing and non-desorbing fractions of phenanthrene and pyrene in four test non-sterilized soils as a function of aging time were shown in Figure 1-7. The initial soil (time 0) showed higher available fractions for phenanthrene and pyrene, and the concentrations then generally decreased with time. For example, the concentrations of phenanthrene in soil 1, soil 2, soil 3, and soil 4 were reduced to 24.5 mg/kg, 9.2 mg/kg, 10.9 mg/kg, and 5.6 mg/kg within 16 weeks, while those of pyrene decreased to 60.0 mg/kg, 41.0 mg/kg, 23.7 mg/kg, and 10.1 mg/kg, respectively. The data shown in Figure 1-7 suggest that the aging process significantly affects the availability of the examined PAHs in soils. These PAHs were more readily available at the start of aging, and their availabilities decreased rapidly with increasing the soil-PAH contact time.

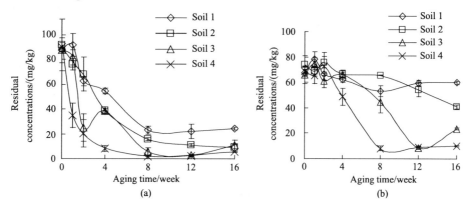

Figure 1-7   Concentrations of available fraction of phenanthrene (a) and pyrene (b) in test non-sterilized soils as a function of aging time

The degradation efficiency of available PAHs in non-sterilized soils was generally higher for those PAHs with low molecular weight, basing on the observation given in Figure 1-7. Significant losses were observed in soil 1, soil 2, soil 3, and soil 4 after 16 weeks of aging, and 72.3%, 90.0%, 87.6%, and 93.8% of available phenanthrene and 11.6%, 44.6%, 64.0%, and 85.2% of available pyrene disappeared in these soils, respectively. Indeed, the available fractions of phenanthrene were degraded to a constant level after 8 weeks of aging in the four test soils, while a similar phenomenon was only observed for pyrene in soil 1 and soil 4. In the same soil samples, the available phenanthrene was more readily degradable and the dissipation ratio of the available phenanthrene was clearly higher than those of pyrene, indicating that PAHs with higher molecular weight and more benzene rings were more recalcitrant in soils. This was also supported by previous findings (Horinouchi et al., 2000; Sabaté et al., 2006; Derudi et al., 2007).

The desorbing fraction was the predominant portion of the available fractions of the test PAHs (Table 1-3). However, one notes that the ratios of the dissipation amount of the desorbing fraction of pyrene relative to the total dissipation amount of the available fraction in soils 1 and 4 after 16 weeks were greater than 100%, while dissipation of the non-desorbing fraction of pyrene contributed negatively to the total dissipation of its available fraction in soils (Table 1-3). These observations suggest that the forms of PAHs may change over the aging period, and parts of the desorbing fraction of PAHs may transform into the corresponding non-desorbing fraction in soils.

Table 1-3 Ratios of the dissipation amount of the desorbing or non-desorbing fraction of test PAHs relative to the total dissipation amount of their available fraction in soils after 16 weeks

Soil	Phenanthrene		Pyrene	
	D/%	ND/%	D/%	ND/%
Soil 1	98.2±2.1	1.8±0.6	120.1±2.3	−20.1±9.9
Soil 2	97.4±2.8	2.6±0.1	90.2±3.5	9.8±6.0
Soil 3	92.1±5.3	7.9±0.3	76.8±3.6	23.2±7.6
Soil 4	99.2±0.6	0.8±0.2	102.2±6.4	−2.2±1.2

Note: D means the ratio of the dissipation amount of the desorbing fraction of tested PAH to its total dissipation amount of the available fraction in soils after 16 weeks; ND means the ratio of the dissipation amount of the non-desorbing fraction of tested PAH to its total dissipation amount of the available fraction in soils after 16 weeks.

Availability of an organic contaminant in soil environment is controlled by the physicochemical properties of the contaminant and soil. Availability has been shown to vary with soil type, as illustrated by differing pesticide toxicities in various soil types. It is reasonable to assume the influenced availability by soil organic matter (SOM), since partition and therefore aging is highly affected by SOM. Nam et al. (1998) reported the putative correlations between SOM and availability of phenanthrene in soil, and observed that the aging of phenanthrene, as measured by declining availability to phenanthren-degrading microorganisms, was only appreciable in soils with >2.0% organic carbon. However, Chung and Alexander (2002) attempted later to correlate availability of phenanthrene and atrazine with a number of soil properties including SOM, and found that it could not in all cases be predicted by any one soil property or combination of properties. Similarly, in our work, no significant correlations of the available portions of the test PAHs in four soils were observed with SOM and soil textures.

1.2.2 Microbial degradation of available fractions of PAHs in soils

Decreases in the available PAH fractions accessible for microbial degradation commensurate with increasing aging time were observed in the test soils (Figure 1-7). Similar findings indicating decreases in fractions of organic compounds available for biodegradation with aging time have been reported previously (Smith et al., 1999; Alexander, 2000). Here, the availability of phenanthrene in soils decreased rapidly even at the beginning of aging, suggesting the rapid establishment of a potent phenanthrene degrading microbial consortium in soils. However, due to the higher molecular weight and larger number of benzene rings, pyrene was more recalcitrant, and it may take longer ($\geqslant$4 weeks) for the establishment of its microbial consortium in soils. Microbial catabolism has been reported in many studies, and has been described as a principal mechanism involved in the removal of available contaminants such as PAHs from soil (Semple et al., 2003).

In fact, the desorbing fraction of organic chemicals in soils was more readily biodegradable than the non-desorbing fractions. The results of this study indicate that attenuation of the desorbing fraction primarily accounted for the observed reduction in available fractions of PAH over time (Table 1-3). For example, the loss of the desorbing fractions of phenanthrene in the four test soils after 16 weeks contributed 98.2%, 97.4%, 92.1%, and 99.2% to the corresponding total dissipation amounts of available fractions of this compound, respectively. Similarly, the dissipation amounts of desorbing fractions of pyrene also accounted for >76.8% of the total dissipation amounts of its available fractions in soils. The dissipation of available PAH fractions from the non-sterilized soils were primarily due to intrinsic microbial degradation. As an example, the residual concentrations of the available fraction, desorbing fraction, and non-desorbing fraction of phenanthrene were 81.3 mg/kg, 71.7 mg/kg, and 9.6 mg/kg in microbe-inhibited control soils after 16 weeks of aging, but only 10.9 mg/kg, 9.4 mg/kg, and 1.5 mg/kg in non-inhibited soils, respectively. Similar results were observed for pyrene (Figure 1-8). In comparison with the microbial inhibited control soils, the overwhelmingly larger dissipation of extractable residues in non-sterilized soils indicated that microbial degradation was the dominant factor responsible for dissipation of the available fractions of PAHs in soil.

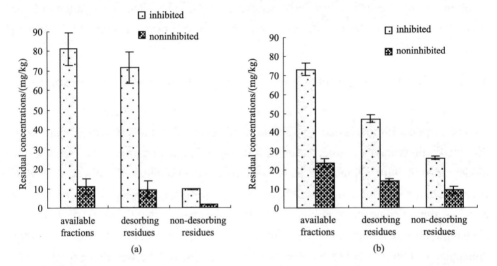

Figure 1-8 Residual concentrations of phenanthrene (a) and pyrene (b) in microbe inhibited and non-inhibited soil 3 aging for 16 weeks

Previous studies have indicated that most heterotrophic bacteria and fungi favor a soil pH of approximately 7, with acidic or basic conditions negatively influencing the ability of microbial populations to degrade PAHs (Leahy and Colwell, 1990; Kastner et al., 1998). In the present study, the degradation ratios of PAHs in soil 1 (pH 4.56) and soil 2 (pH 4.74) were much lower than those in soil 3 (pH 6.02) and soil 4 (pH 7.35) (Figure 1-7). In addition, the PAH degradation ratios were of the same order as the pH values of these soils. Similar results have also been reported by other researchers (Swindell and Reid, 2006). These findings further confirm that the degradation of available PAHs in soils depends markedly on microbial activities in the soil environment.

1.2.3 Transformation of available fraction of PAHs to bound residue in soils

It is generally recognized that as contact time (so called aging) between organic contaminants and soil increases, the contaminant molecules can slowly diffuse into trapped sites in soil organic matter and mineral-organic complex, resulting in the phenomenon of "contaminant's sequestration" (Pignatello and Xing, 1996; Nam et al., 1998; Chung and Alexander, 2002; Gunasekara et al., 2003). The sequestrated contaminant would ideally transformed to bound residues and reduced its availability in soil with prolonged aging time.

The forms of HOCs in the soil environment have been reported previously (Loiseau and Barriuso, 2002; Lesan and Bhandari, 2004), and the formation of bound residues would theoretically reduce the availabilities of HOCs in soils. However, previous studies have indicated that the formation of bound residues of HOCs was negligible in soil (Sabaté et al., 2006). In this study, bound residual phenanthrene and pyrene were detected in the four test soils (Table 1-4). The bound residual PAH concentrations tended to first increase and then decrease over the 16-week aging period. However, in comparison with the available PAH residues, the concentrations of bound PAH residues were always very low with maximum values of 0.60 mg/kg and 1.47 mg/kg for phenanthrene and pyrene in soil 1, respectively (Table 1-4). Although the presence of microbes resulted in partial transformation from available fractions to bound residues (Richnow et al., 2000), the results of the present study confirm that the contribution of this process to the dissipation of available fractions of PAHs in soils was negligible.

Table 1-4 Concentrations of bound residue of phenanthrene and pyrene in soil 1

Time/week	Phenanthrene/ (mg/kg)	Pyrene/(mg/kg)
0	0.097±0.13	0.21±0.098
1	0.14±0.59	0.38±0.11
2	0.39±0.16	0.58±0.28
4	0.60±0.23	1.47±1.11
8	0.20±0.17	0.42±0.31
12	0.34±0.15	0.53±0.40
16	0.078±0	0.63±0

1.2.4 Butanol-extraction technique for predicting the availability of PAHs in soil

Assessment of the bioavailability of organic contaminants in soil is considered to be important in understanding the risk posed by the contaminant and the means required for successful bioremediation (Alexander, 2000; Semple et al., 2003). Much evidence exists that vigorous extraction may overestimate the exposure of various organisms to organic pollutants in soil, because it often overestimates the bioavailable concentrations (Tang et al., 2002; Ling et al., 2009). It is thus important to have methods to assess the actually bioavailable concentrations of these compounds. Basing on the principle that microbial bioavailability is governed by chemical mass transfer from the soil solid phase to the aqueous phase, a number of non-exhaustive extraction techniques have been proposed including solid phase extraction, supercritical fluid extraction, surfactant extraction, persulfate oxidation, solvent extraction, and cyclodextrin extraction (Swindell and Reid, 2006). Kelsey et al. (1997) firstly reported that *n*-butanol was a most appropriate extraction solvent for predicting bioavailability of phenanthrene to earthworms and bacteria. Liste and Alexander (2002) further proved that the quantities of phenanthrene and pyrene biodegraded in soil significantly correlated with their amounts extracted by *n*-butanol, and the amount of PAHs removed by six consecutive groups of earthworms was similar to the quantity extracted by butanol. As such, in many studies, the *n*-butanol extraction method was also used to evaluate the availabilities of PAHs in soil (Ling et al., 2009).

1.2.5 Phytoavailability of bound-PAH residues in soils

The formation of a bound residue is considered to act as a soil detoxification process by permanently binding compounds into soil matrices, and the bioavailability of bound residues is the final endpoint for the risk assessment and regulatory management of organic chemicals in the soil environment (Richnow et al., 1995; Northcott and Jones, 2000). However, in recent decades, some research has also suggested that bound residues should not be considered permanently bound and that they pose a risk to human health since there exists the potential exists for partial reversibility and subsequent availability (Northcott and Jones, 2000; Käcker et al., 2002; Barraclough et al., 2005). However, despite numerous studies of bound residues using radioactive tracer compounds, the majority of their chemical properties remain unknown (Nowak et al., 2011). To our knowledge, little information is currently available on the phytoavailability of bound-PAH residues in soils.

In this work, we firstly investigated the phytoavailability of bound-PAH residues in soils (Gao et al., 2013). Soil 3 in Table 1-2 was sieved through a 2-mm mesh, and then PAH-spiked and aged for 45 days at 22 ℃ in the dark with a water content of 20% of the soil water-holding capacity. Soil samples were collected, freeze-dried, and solutions of dichloromethane:acetone (1 ∶ 1, vol/vol) were added (1 g soil with 5 mL solution). Extraction was conducted in an ultrasonic bath for 10 min. The solvent was decanted, and the samples were then re-extracted with the replenished fresh solvent and sonicated. This process was repeated six times, and then soils were air-dried and sieved through a 2-mm mesh (Sabaté et al., 2006). The soils contained only bound residues of PAHs were obtained. The final concentrations of bound residual phenanthrene and pyrene in soils were 0 mg/kg to 2.23 mg/kg and 0 mg/kg to 6.91 mg/kg, respectively. Then the soils packed into greenhouse pots, and ryegrass (*Lolium multiflorum* Lam.) was planted. Plants were destructively sampled 45 days after sowing, and used for PAH analysis.

Although the test soils contained only bound residues of phenanthrene and pyrene, the obvious uptake, accumulation and translocation of phenanthrene and pyrene by ryegrass indicated that these bound residual fractions were significantly phytoavailable in the soil environment.

The concentrations of test PAHs, on a dry weight basis, in ryegrass roots and shoots as a function of their residual concentrations in soils are shown in Figure 1-9, which indicates that root concentrations of phenanthrene and pyrene were monotonically enhanced by increasing their soil concentrations. With the increase of soil phenanthrene and pyrene concentrations (from 0 mg/kg to 1.10 mg/kg and from 0 mg/kg to 4.09 mg/kg, respectively) phenanthrene and pyrene concentrations in the roots of ryegrass increased from 0 mg/kg to 14.1 mg/kg and from 0 mg/kg to 29.0 mg/kg, respectively. Because the concentrations of test PAHs in roots grown in PAH-free control soils were not detectable, the conclusion was made that the PAHs in roots were derived from root uptake from the surrounding soils.

In addition, concentrations of phenanthrene and pyrene in the shoots of ryegrass also generally increased with an increase in their respective soil concentrations (Figure 1-9). However, the PAH concentrations in shoots were substantially lower than those in roots for the same soil-plant treatment, implying that the translocation of phenanthrene and pyrene from roots to shoots was markedly restricted. Note that the shoot concentrations of test PAHs were also undetectable in the unspiked control soils, although all treated plants shared the same atmospheric conditions. This differed from results previously reported using freshly spiked soils in which even plants grown in unspiked control soils had shoots that accumulated PAHs from the atmosphere via PAH evaporation from soils (Zhu and Gao, 2004; Gao et al., 2008).

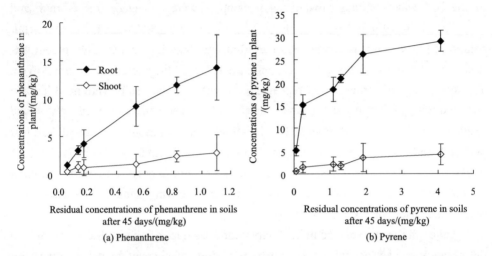

(a) Phenanthrene   (b) Pyrene

Figure 1-9 Concentrations of phenanthrene and pyrene in plant roots and shoots as a function of their concentrations in soils. Error bars represent standard deviations (SD)

Plant concentration factors (PCFs) were recorded as the ratio of the phenanthrene or pyrene concentration in a plant (C_p, on a dry weight basis) or plant part to the concentration in the soil (C_s) when sampling, that is, PCF = C_p/C_s (Burken and Schnoor, 1998; Li et al., 2002; Gao et al., 2008b). Root concentration factors (RCFs) and shoot concentration factors (SCFs) for the uptake of phenanthrene and pyrene by ryegrass as a function of their soil concentrations after 45 days are shown in Figure 1-10. Both RCFs and SCFs generally decreased with an increase in the PAH concentrations in soils. The RCFs of phenanthrene and pyrene within the test soils were 12.8 L/kg to 24.6 L/kg and 7.1 L/kg to 94.9 L/kg, respectively. The SCFs of phenanthrene and pyrene were 2.4 L/kg to 7.7 L/kg and 1.0 L/kg to 10.4 L/kg, respectively; generally much smaller than the RCFs for the same plant-soil treatment. In addition, the RCFs for phenanthrene uptake were always much smaller than those for pyrene uptake by ryegrass when both PAHs were at the same concentration in the soil.

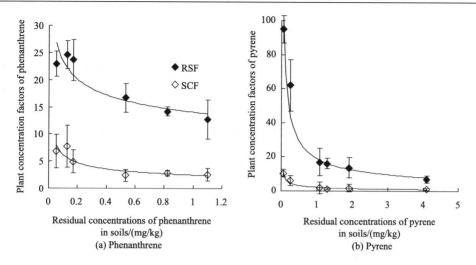

Figure 1-10 Root concentration factors (RCFs) and shoot concentration factors (SCFs) for ryegrass uptake of (a) phenanthrene and (b) pyrene in soils. Error bars represent standard deviations (SD)

The translocation factor (TF) was used to evaluate the translocation of PAHs from plant roots to shoots. The TF values were calculated as TF = SCF/RCF (Mattina et al., 2003; Gao et al., 2008b). As shown in Figure 1-11, the average TF values for the uptake of phenanthrene and pyrene by ryegrass from test soils were only 0.23 and 0.11, respectively, indicating a restriction in their translocation from root to shoot. Moreover, TFs for phenanthrene were always much larger than those for pyrene.

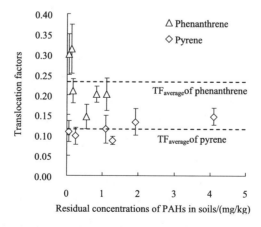

Figure 1-11 Translocation factors (TFs) for ryegrass uptake of phenanthrene and pyrene in soils. Error bars represent standard deviations (SD)

Laboratory studies have shown that bound PAH residue formation is a primary fate mechanism of PAHs. Bound residues have a direct effect on the long-term partitioning behavior, bioavailability, and toxicity of organic compounds in soil (Suflita and Bollag, 1981; Pignatello, 1998; Northcott and Jones, 2000). They represent compounds in the form of the parent substance or its metabolites after extraction. In general, the formation of bound residues significantly reduces the bioaccessibility and bioavailability of bound residues and is the final endpoint for the risk assessment and regulatory management of organic compounds in the soil environment (Führ et al., 1998). Richnow et al. (1999) documented that on anthracene immobilization within bound residues, the bioavailability of residual anthracene fragments decreases by an order of magnitude as compared to the parent anthracene. Guthrie and Pfaender (1998) showed that ^{14}C residues of pyrene associated with humic/fulvic acid extracts and residual soil after acid digestion were not readily bioavailable to a pyrene-mineralizing microbial community. Eschenbach et al. (1998) documented that bound-PAH residues were degraded at limited rates, similar to the natural humus turnover rate, and that the residue extractability did not increase after exposure to extreme environmental changes such as freezing/thawing or wetting/drying cycles. However, this investigation clearly revealed that the bound-PAH residues in soils remained phytoavailable and degradable, posing serious threats to human health and ecological safety.

Some studies have concluded that bound residues should not be considered permanently bound as the potential for partial reversibility always exists, and pollutants may subsequently be released from soil by the continuous turnover of soil organic carbon or undergo further entrapment and/or binding as a result of humification and diagenesis (Northcott and Jones, 2000). Evidence further suggests that bound residues can be remobilized into solution or undergo further alteration by microbial interaction and chemical reaction (Launen et al., 2000). Mechanisms of compound sequestration such as intraorganic matter diffusion (IOMD) and intraparticulate diffusion (IPD) retard the release of sorbed compounds, but SOM and environmental particles are themselves subject to alteration by environmental, chemical, and biotic processes (Käcker et al., 2002). Natural soil humification processes will continue to degrade the macromolecular humic polymers and may release previously bound residues with half-lives equivalent to those for humic materials. Biota residing in soils can promote bound residue release by digestive extraction and excretion, enhancing enzymatic release by extracellular excretions, SOM metabolism, and bioturbation (Launen et al., 2000).

Here, the phytoavailability of the bound-PAH residues in soil were indicative of their significant release and subsequent uptake by plants and/or degradation in the rhizosphere soil (Gao et al., 2013). This was largely ascribed to the enhanced availability of bound-PAH residues due to the input of root exudates to the test soils, which will be elucidated in the following chapters.

Chapter 2 Gradient distribution of PAHs in rhizosphere soil

Toxic substances enter plant roots through the rhizosphere (Gao et al., 2011). During plant growth, roots actively or passively release organic compounds, referred to as root exudates (Phillips et al., 2003; Gao et al., 2010). The rhizosphere is defined as the volume of soil shared with soil bacteria and over which roots have an influence (Joner and Leyval, 2003; Ling et al., 2013). The primary drivers of rhizosphere formation are the development of water and solute gradients around roots, which can alter the physical, chemical, and biological properties of soil. Up to 40% of net carbon fixed by plant shoots during photosynthesis is released as root exudates into soils (Lynch and Whipps, 1990). Rhizodegradation of organic pollutants may be improved by root exudates, which affect rhizospheric microbial processes (White et al., 2003; Corgié et al., 2004). The elevated degradation of organic pollutants in the rhizosphere has been well documented when compared to bulk soil (Joner et al., 2001). The distribution of organic contaminants in the rhizosphere is clearly paramount to their fate in soil-plant system.

After diffusion into rhizosphere soil, soil near the rhizosphere zone, or bulk soil, root exudates gradually disappear due to radial dilution and microbial consumption (Corgié et al., 2003; He et al., 2009; Gao et al., 2011). This may result in the gradient distribution of root-derived substances, and possibly generate a gradient of PAH-degradation between the rhizosphere and bulk soil.

Rhizodegradation of recalcitrant organic pollutants may benefit from the major role that root exudates have on rhizospheric microbial processes (Jones et al., 2001; Singh et al., 2004). Because these root exudates are a convenient source of carbon and energy, they may represent an important resource for fast-growing microbes and consequently alter the species composition of the rhizosphere, which functions in nutrient transformation, decomposition, and mineralization of organic substances (Paterson et al., 2000; Smalla et al., 2001; Costa et al., 2006). Because production of rhizodeposits can vary during plant and root development, microbial communities in the rhizosphere are influenced by the developmental stage and location in particular parts of the root system (Yang and Crowley, 2000). A bacterial gradient was observed, with higher numbers of heterotrophs and PAH-degrading bacteria closest to the roots (Corgié et al., 2003). Later, using direct DNA extraction techniques, PCR amplification, and thermal gradient gel electrophoresis screening, Corgié et al. (2004) further observed that bacterial communities in three layers (0~3 mm, 3~6 mm, and 6~9 mm from the root mat) of the rhizosphere differed in both the presence and absence of phenanthrene as a function of distance from the roots. However, most previous studies focused on the rhizosphere microbe community (Smalla et al., 2001; Nunan et al., 2005). Few reports have addressed the gradient distribution of recalcitrant organic pollutants such as PAHs in the rhizosphere.

2.1 Gradient distribution of PAHs in rhizosphere soil: a greenhouse experiment

Here, the gradient of phenanthrene and pyrene as representative PAHs was investigated in rhizosphere soil in proximity to the root surface using a greenhouse pot experiment. Pot experiments have commonly been used to investigate the mechanisms, fate, transport, and transformation of organics in the rhizosphere because potted soils are more homogeneous, and pot studies can be conducted under well controlled conditions (Binet et al., 2000).

Soil 3 in Table 1-2 was used. Soil samples were air-dried and sieved through 2-mm mesh. Soil was spiked with PAHs in acetone to create polluted soil (Gao et al., 2011), and the final concentrations of phenanthrene and pyrene in soil of 250 mg/kg and 100 mg/kg, respectively. Ryegrass (*Lolium multiflorum* Lam.) was vegetated in greenhouse pots with contaminated soils. The treated pots were kept in the greenhouse at 25~30°C during daytime and 20~25°C during the night and relocated randomly every 2 days. Plant and soil were sampled at 40-day and 50-day after sowing (Gao et al., 2011).

The rhizosphere soils in proximity to the root surface were sampled according to previous studies (Joner and Leyval, 2003; Ling et al., 2013). The three layers of rhizosphere soil from the ryegrass root surface were divided into rhizoplane, strongly adhering, and loosely adhering soil. Assuming that these portions of the soil had formed consecutive cylindrical layers around the roots, an average thickness of each layer was estimated based on soil mass and density, mean root diameter, and total root length in each pot. At harvest, shoots were cut at the soil surface and weighed. The upper 2~5 mm of soil in each pot and the soil that did not adhere to roots was discarded.

Roots and remaining soil were further separated by gently crushing the soil and shaking out the roots, and the portion of soil obtained in this manner was classified as loosely attached to roots (average 8 mm proximity to root). Soil that required continued vigorous rubbing and shaking of the root system was classified as strongly attached to roots (average 4 mm proximity to root). After this sequential soil separation by shaking, the intact root system was washed in a large beaker with 0.5 L deionized water, and the adhering soil was recovered on a Whatman GF/A glass microfiber filter; this fraction was classified as rhizoplane soil (average 1 mm proximity to root). Then samples of rhizoplane, strongly adhering, and loosely adhering soil were analyzed for PAH concentrations.

2.1.1 Gradient distribution of phenanthrene and pyrene in rhizosphere

Three portions of soil, loosely adhering, strongly adhering, and rhizoplane soil, were sampled, corresponding to an average of 8 mm, 4 mm, and 1 mm from the root surface of ryegrass. The soil concentrations of phenanthrene and pyrene were determined (Figure 2-1) after 40 days or 50 days of plant growth, and the detected PAH

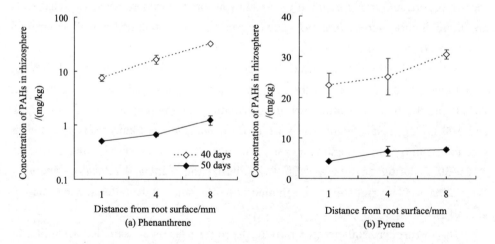

Figure 2-1 Gradient concentrations of phenanthrene (a) and pyrene (b) in non-sterilized rhizosphere soil after 40 days and 50 days as a function of the distance from root surface. The initial concentrations of phenanthrene and pyrene in spiked soils were 250 mg/kg and 100 mg/kg, respectively. Error bars are standard deviation (SD)

concentrations clearly increased in soil with distance from the root surface. Concentrations of phenanthrene and pyrene in loosely adhering soil were 34% and 33% higher, respectively, than those in rhizoplane soil after 40 days, and 150% and 71% higher after 50 days. Compared with the 50-day treatment, PAH concentrations were generally much higher in soils after 40 days than after 50 days, indicating obvious degradation of PAHs in all three layers of rhizosphere soil.

The degradation ratios (D) of PAHs in each layer were calculated according to the equation:

$$D\,(\%) = (C_o - C_s)/C_o \times 100\%$$

where C_o is the initial concentrations of PAHs in soil, and C_s is the residual concentrations of test PAHs in soil after cultivation. 87%~97% of phenanthrene and 69%~79% of pyrene dissipated in the three layers of the rhizosphere soil after 40 days, and more than 99% of phenanthrene and 93% of pyrene dissipated in these areas after 50 days. The degradation ratios differed significantly between PAHs. The D values of phenanthrene in each fraction of soil were always higher than those of pyrene, irrespective of cultivation time. Additionally, the D values clearly decreased in soils with distance from the root surface, i.e., selected PAHs in rhizoplane soils were more readily degraded than those in loosely or strongly adhering soils.

Among different PAHs, phenanthrene seems to be more readily degradable, and its degradation ratio in each layer of the rhizosphere was correspondingly higher than that of pyrene (Figure 2-1). Phenanthrene, a low-molecular-weight PAH with three aromatic rings, is metabolized by many different microorganisms, and the presence of a complex mixture of microorganisms results in the most efficient biodegradation of phenanthrene (Derudi et al., 2007). The above results indicate that PAHs of higher molecular weight and with more benzene rings are more recalcitrant in soils, as supported by previous findings (Ling et al., 2010; Sabaté et al., 2006).

2.1.2 Gradient distribution of root exudates in rhizosphere

Ryegrass was used to obtain high root exudation, as this plant is reported to release high amounts of exudates into the rhizosphere (Corgié et al., 2003). Grasses are commonly used in phytoremediation due to their fibrous root system and large specific root surface area. They exude high quantities of soluble organic substrates, up to 500 mg/g root dry weight after 4 weeks of growth following germination (Lynch and Whipps, 1990), which may represent 65% of net CO_2 assimilated by ryegrass and translocated into the rhizosphere in 4 weeks (Meharg and Killham, 1990). Moreover, roots of ryegrass have a short life span, with only 50% surviving more than 30 days (Forbes et al., 1997). Hence, ryegrass was chosen as a test plant in this investigation.

Root exudates in each layer of the rhizosphere soil were examined in this study. Soluble organic carbon (SOC) is commonly used to characterize the total amount of root exudates. Organic acids (OA) and total soluble sugar (TSS) are two important components of low-molecular-weight root exudates (Xie et al., 2008). Although many kinds of organic acid are secreted from roots into the soil environment, oxalic acid is the major acid secreted. Our previous work showed that oxalic acid was the main component of low-molecular-weight organic acids of ryegrass root exudates and constituted about 90% of total organic acids (Yang et al., 2010; Gao et al., 2011), as also reported by Xie et al. (2008). Hence, oxalic acid was selected as a representative organic acid and, together with SOC and TSS, was utilized to characterize the distribution of root exudates in PAH-contaminated rhizosphere soils.

The gradient distribution of detected root exudates in the three layers of non-sterilized rhizosphere soil is shown in Figure 2-2. The concentrations of SOC, OA and TSS in PAH-spiked soils clearly decreased with distance from the root surface; their concentrations in rhizoplane soil were 235%, 86.2% and 141% higher, respectively, than in loosely adhering soil after 40 days, and they were 64.1%, 81.5% and 190% higher after 50 days. Furthermore, root exudates tended to accumulate in each layer of the rhizosphere with time, and the concentrations of SOC, OA and TSS were clearly higher after the 50-day than after the 40-day treatment.

It appears that the concentration of root exudates in each layer of rhizosphere soil depends mainly on two aspects: root exudation and diffusion and degradation due to chemical and/or biological processes. Away from the root surface, root exudates diffuse radially into layers of the rhizoplane soil, strongly adhering soil, and loosely adhering soil. However, a portion of the root exudates is degraded in soils during the diffusion process, resulting in a decrease in concentrations with distance from the roots.

Table 2-1 provides the gradient distribution of root exudates in sterilized rhizosphere soil close to the root surface. Compared with non-sterilized soil (Figure 2-2), a similar trend was observed with a decreased gradient of root exudates in each of the three layers of sterilized rhizosphere soil from the root surface, but the observed concentrations of SOC, OA and TSS were correspondingly 120%~232%, 86%~187% and 47%~137% higher than those in non-sterilized soil treatments. This finding was consistent with the result that microbial consumption greatly contributed to the dissipation of root exudates.

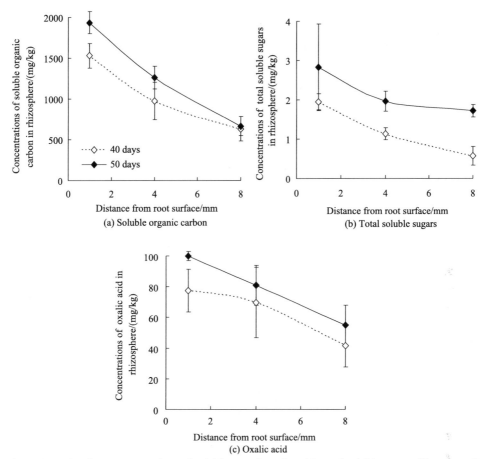

Figure 2-2 Gradient concentrations of soluble organic carbon (a), total soluble sugars (b), and oxalic acid (c) in non-sterilized rhizosphere soils after 40 days and 50 days as a function of the distance from root surface. The initial concentrations of phenanthrene and pyrene in soils were 250 mg/kg and 100 mg/kg, respectively. Error bars are standard deviation (SD)

Table 2-1 Gradient concentrations of root exudates in sterilized PAH-spiked soils after 50 days(mg/kg)

Root exudates	Averaged distance in proximity to root surface		
	1 mm	4 mm	8 mm
Soluble organic carbon	4262 (869.3)a	3424 (1001)ab	2215 (254.5)b
Soluble carbohydrate	6.693 (0.222)a	4.034 (0.102)b	2.522 (0.813)c
Oxalic acid	185.7 (9.611)a	156.2 (30.23)ab	157.7 (17.53)b

Note: The initial concentrations of phenanthrene and pyrene in spiked soils were 250 mg/kg and 100 mg/kg, respectively. Values in the same line followed by the same letter are not significantly different ($p < 0.05$). Data in bracket are standard deviation (SD).

The higher concentrations of root exudates in sterilized versus non-sterilized soils indicate that microbial consumption greatly contributed to the dissipation of root exudates. Root exudates provide the necessary nutrition and energy for survival and reproduction of rhizosphere microbes (Overbeek et al., 1995; Binet et al., 2000; Joner and Leyal, 2003; Gao and Zhu, 2005). In sterilized soil, microbial activity and reproduction are limited. Due to microbial activity, less root exudate was degraded in the sterilized treatment compared with the non-sterilized treatment, theoretically resulting in higher exudate concentrations in the sterilized rhizosphere.

Notably, the detected concentrations of root exudates are actually the net result of root exudation and biological and chemical consumption. A limited number of methods are available to determine the actual amount of root exudates secreted by plants in the rhizosphere, and little information is available on the detailed processes of microbial consumption and reproduction of root exudates. Although the chemical behaviors (such as sorption and desorption) of several root exudate fractions including organic acids, amino acids, and aromatics in soil have been reported (Gao et al., 2003; van Hees et al., 2005), the actual fixed and degraded amounts of root exudates in the rhizosphere due to chemical processes are still difficult to detect.

2.1.3 The correlations of PAH concentration gradient with the concentration gradient of root exudates in rhizosphere

Comparing Figures 2-1 and 2-2 shows that the concentration gradient of detected root exudates was significantly and negatively correlated with the PAH concentration gradient and positively correlated with the PAH degradation ratio in the rhizosphere (as shown in Figures 2-3 and 2-4), indicating the major role of root exudates in PAH dissipation.

Rhizoremediation of organic pollutants, including PAHs, depends predominantly on rhizosphere-stimulated microbial biodegradation, and plant uptake contributes negligibly to their dissipation in soil (Gao and Ling, 2006). Root exudates enhance microbial biodegradation in the rhizosphere in three ways: direct degradation by enzyme systems in root exudates, improved living conditions for growth and activity of indigenous microorganisms, and as a carbon source for microbial biodegradation of organic chemicals. Several enzymes secreted by plants, such as peroxidases and phenol oxidases, have been shown to degrade aromatic hydrocarbons (Lu et al., 2008).

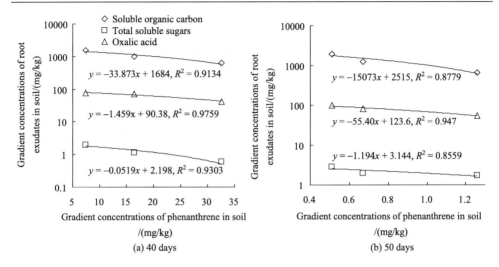

Figure 2-3 The correlation of the gradient in phenanthrene concentration with the gradient in concentration of root exudates in the non-sterilized rhizosphere soils close to the root surface (0~8 mm) after 40 days (a) and 50 days (b), respectively

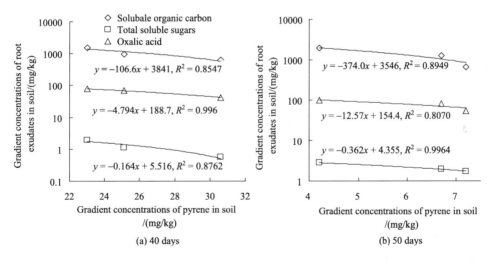

Figure 2-4 The correlation of the gradient in pyrene concentration with the gradient in concentration of root exudates in the non-sterilized rhizosphere soils close to the root surface (0~8 mm) after 40 days (a) and 50 days (b), respectively

It is generally recognized that large numbers of microorganisms with high activity in the rhizosphere dominate the rhizodegradation of organic contaminants in soil (Binet et al., 2000; Gao and Zhu, 2005). The number of microorganisms and their activity in the rhizosphere are much higher than that in bulk soil, and the release of root exudates plays significant roles in rhizosphere formation (Johnson et al., 2005; Rentz et al., 2005). Additionally, the composition and quantity of root exudates have been demonstrated to affect rhizosphere microbial flora (Yang and Crowley, 2000; Jones et al., 2004; Singh et al., 2007). Utilizing special equipment, Yoshitomi and Shann (2001) directly observed the accelerated degradation of ^{14}C-pyrene in soil by adding actual corn (*Zea mays* L.) root exudates. Similarly, Rentz et al., (2005) reported that root exudates of white mulberry (*Morus alba* L.) dramatically helped remove benzo[a]pyrene, a PAH with a higher molecular weight, by co-metabolism. These observations indicate that the presence of root exudates in the rhizosphere enriches the microorganisms and thus enhances rhizodegradation of organic contaminants in the soil environment.

However, most previous studies focused on the rhizosphere microbe community (Reilley et al., 1996; Nunan et al., 2005), and few reports have documented the gradient distribution of microbes in soils close to the root surface. We postulated that larger amounts of root exudates in soil near the root surface mean that larger quantities of microorganisms with higher microbial activity would be available, which, in turn, would stimulate the degradation of PAHs in the rhizosphere (Figure 2-1). This observation was also supported in the study by Corgié et al., (2003), who used a root-box device and observed a bacterial gradient, with higher numbers of heterotrophs and PAH-degrading bacteria in the PAH-spiked rhizosphere closest to the roots (from 0 mm to 9 mm). Similarly, the observed increase in PAH concentrations and decrease in their degradation ratio in the rhizosphere with increasing distance from the root surface could be mainly due to the decreased gradient in amounts and activities of microorganisms in these areas.

Root exudates not only serve as convenient sources of carbon and energy for microorganisms, but also affect the availability of recalcitrant organic pollutants, including PAHs, in the rhizosphere (Ouvrard et al., 2006; Ling et al., 2009; Gao et al., 2010). Utilizing laboratory batch assays, we revealed that 0~1000 mmol/L low-molecular weight organic acids, citric acid and oxalic acid as representatives, significantly inhibited sorption and promoted desorption of phenanthrene in soil (Gao et al., 2010b). Microcosm experiments showed that the *n*-butanol extractable amounts of phenanthrene and pyrene in soils increased with increasing the concentrations of citric and oxalic acid from 0 g/kg to 57.6 g/kg and from 0 g/kg to 27.0 g/kg, respectively, and that adding citric acid promoted phenanthrene and pyrene availability to a greater degree than did adding oxalic acid (Ling et al., 2009). The enhanced availability of PAHs in soils by complex root exudates was observed recently in our work (Gao et al., 2010). Yang et al.(2001) also reported similar phenomena, namely that some root exudates, such as citric acid, increased the availability of weathered PAHs in soils. Here, the test PAHs in layers of the rhizosphere near roots may be more available and readily biodegradable due to the presence of higher amounts of root exudates, which may have contributed to the increased degradation ratios of PAHs in these areas.

One notes that, in this investigation, loosely adhering, strongly adhering, and rhizoplane soil were collected as referred to in the literature (Joner and Leyval, 2003), assuming that these rhizosphere soils had formed consecutive cylindrical layers around roots. The average thickness of each layer was estimated based on soil mass and density, mean root diameter, and the total root length in each pot. Although this method is effective for rhizodistribution investigations of organic pollutants, the collection processes did not seem sufficiently rigorous, and the calculated distance of the rhizosphere soil from the root surface might not be perfectly consistent with the actual environment. However, a limited number of methods to identify the actual conditions in the different cylindrical layers of rhizosphere soil around roots are available (Corgié et al., 2003).

2.2 *In situ* gradient distribution of PAHs in rhizosphere soil: a field study

Based on greenhouse pot experiments, a bacterial gradient was observed by Corgié et al. (2003), with an increased number of heterotrophs and PAH-degrading bacteria closest to the roots. We found that the residual concentrations of PAHs increased from the rhizoplane to loosely adhering soil after 40 days and 50 days, which was significantly and negatively correlated with the amount of root exudate in the rhizosphere (Gao et al., 2011). Similar findings were reported by Joner and Leyval (2003) and Corgié et al. (2004). However, the few studies on gradient rhizodegradation and PAH distribution in the rhizosphere were investigated using pot experiments. Field research would be ideal to determine the biological contaminant distribution in the rhizosphere. In this investigation, we further determined the gradient distribution of PAHs in rhizosphere soil on a field scale (Ling et al., 2013).

Contaminated soils were found around a petrochemical plant in Nanjing, China.

The soil type is TypicPaleudalf, a typical zonal soil in East China, with a pH of 5.87, 13.6 g/kg soil organic carbon content, 26.3% clay, 13.0% sand, and 60.7% silt. Clover (*Trifolium pratense* L.) and hyssop (*Hyssopus officinalis* L.) grew *in situ* in this contaminated field soil near a petrochemical plant, and were harvested when about 30 cm tall with mature roots. Rhizosphere soils of the plants were sampled including the rhizoplane, strongly and loosely adhering soil according to the method provided by Joner and Leyval (2003). PAHs were detected in each layer of rhizosphere soils in proximity to the root surface.

2.2.1 *In situ* gradient distribution of PAHs in rhizosphere soil

11 EPA-priority PAHs were detectable in test contaminated soils near a petrochemical plant (Table 2-2). The total PAH concentrations in bulk soil were 95.29 mg/kg, among which 2-ringed and 3-ringed PAHs dominated the 60.88% and 32.23% contribution. By contrast, 5-ringed and 6-ringed PAHs with high molecular weight contributed only 0.86% to the total PAHs in bulk soils.

Table 2-2 Gradient concentration of PAHs in soils proximal to the root surface/(mg/kg) of clover (*Trifolium pratense* L.)

PAHs	Abbreviation	Clover			
		0~3mm	3~6mm	6~9mm	Bulk soil (>9 mm)
Naphthalene	NAP	36.89c	40.72b	41.11b	58.01a
Acenaphthylene	ACE	n.d.	0.04b	0.14a	0.19a
Fluorene	FLU	n.d.	n.d.	n.d.	0.01
Acenaphthene	ACP	3.46c	3.50bc	3.80ab	4.12a
Phenanthrene	PHE	21.15c	21.51bc	23.61ab	25.53a
Anthracene	ANT	0.44c	0.60b	0.73ab	0.86a
Pyrene	PYR	0.12c	0.17bc	0.19b	0.38a
Chrysene	CHR	3.43c	4.24bc	4.46b	5.37a
Benzo(b)fluoranthrene	BBF	n.d.	n.d.	n.d.	0.09
Dibenz(ah)anthracene	DBA	0.44b	0.48b	0.56ab	0.65a
Benzo(ghi)perylene	BGP	n.d.	n.d.	n.d.	0.08
Total PAHs		65.93c	71.26b	74.60b	95.29a

Note: n.d. indicates under the detection limit. Values in the same lines followed by the same letter are not significantly different ($p < 0.05$).

The rhizospheric soil concentrations of PAHs were determined (Table 2-2 and Table 2-3), and total PAH concentrations ranged in 65.93~74.60 mg/kg and

66.42~88.13 mg/kg for clover and hyssop, respectively, which were significantly lower than their concentrations in bulk soils. Similar to the latter, the 2-ringed and 3-ringed PAHs dominated the contribution, while 5-ringed and 6-ringed PAHs contributed negligibly to the total PAH concentrations in rhizosphere.

Table 2-3   Gradient concentration of PAHs in soils proximal to the root surface(mg/kg) of hyssop (*Hyssopus officinalis* L.)

PAHs	Hyssop			
	0~3mm	3~6mm	6~9mm	Bulk soil (>9 mm)
NAP	36.26c	40.63b	54.07a	58.01a
ACE	n.d.	0.16a	0.18a	0.19a
FLU	n.d.	n.d.	n.d.	0.01
ACP	3.36c	3.65b	3.67b	4.12a
PHE	21.49b	22.36b	23.38ab	25.53a
ANT	0.52c	0.63b	0.70b	0.86a
PYR	0.24b	0.25b	0.29b	0.38a
CHR	4.13c	5.02b	5.33ab	5.37a
BBF	0.02c	0.04c	0.07b	0.09a
DBA	0.36b	0.36b	0.37b	0.65a
BGP	0.04b	0.06a	0.07a	0.08a
Total PAHs	66.42c	73.16bc	88.13a	95.29a

Note: n.d. indicates under the detection limit. Values in the same lines followed by the same letter are not significantly different ($p < 0.05$).

Three portions of rhizosphere soil, rhizoplane soil, strongly adhering, and loosely adhering soil, were sampled, corresponding to 0~3 mm, 3~6 mm, and 6~9 mm, respectively, in proximity to the root surface of clover and hyssop. The total PAH concentrations clearly increased in the soil with distance away from the root surface (Figure 2-5(a)). The concentrations of each detectable EPA-priority PAH in three-layered rhizosphere soil were given in Tables 2-2 and 2-3. Similar to the trend of total PAHs, the concentrations of each PAH in soil also followed the descending order of bulk soil, loosely adhering, strongly adhering, and rhizoplane soil. On the other hand, although the chemical properties and soil concentrations varied greatly, the distribution gradient of 2-ringed to 6-ringed PAHs also significantly increased in soils from root surface to the bulk soil, as shown in Figure 2-5(b-f). Results above indicate that rhizosphere effects obviously influence the distribution of PAHs in soil environment.

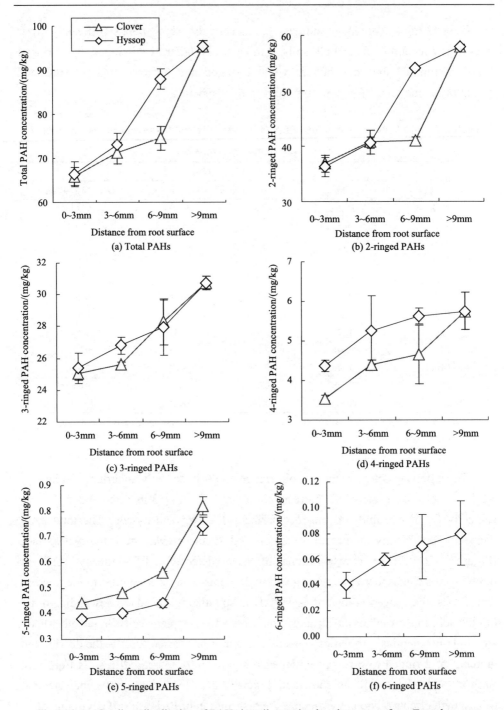

Figure 2-5　Gradient distribution of PAHs in soils proximal to the root surface. Error bars are standard deviation (SD)

2.2.2 Rhizosphere effects on PAH distribution in soil

The rhizosphere effect (R, %) on PAH distribution could be calculated as follows

$$R\ (\%) = (C_{bulk} - C_{rhizo}) / C_{bulk} \times 100\%$$

Where C_{bulk} and C_{rhizo} are the respective PAH concentrations in bulk and rhizosphere soils. The larger R values indicate the more significant rhizosphere effect on PAH distribution in soil. The R values of the total PAHs in proximity to root surface were displayed in Figure 2-6. As seen, R values of total PAHs clearly decreased with the distance away from root surface. A more significant decrease of R values was observed for hyssop versus clover, and the R values of total PAHs in 0~3 mm, 3~6 mm and 6~9 mm rhizospheric layers of hyssop were 30.3%, 23.2% and 7.5%, and those of clover were 30.8%, 25.2% and 21.6%, respectively.

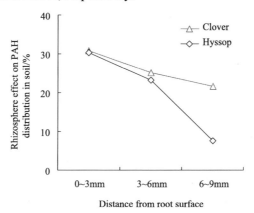

Figure 2-6   Rhizosphere effect (R, %) on the distribution of total PAHs in soil. Error bars are standard deviation (SD)

The R values of different groups of PAHs in rhizosphere soils of clover and hyssop were listed in Table 2-4. Similar to the trend of total PAHs, R values of 2-ringed to 6-ringed PAHs also dropped obviously in three rhizosphere layers along the distance from root surface. The biggest R values were all observed in rhizoplane, and smallest ones in loosely adhering soils irrespective of the PAH properties and plant species. Interestingly, the R values were generally lower for 3-ringed and 4-ringed PAHs in three layers of rhizosphere soils, compared with smaller (2-ringed) and bigger PAHs (5-ringed and 6-ringed). This was more significant in loosely and strongly adhering rhizosphere layers. In addition, as to different plant species, the R values dropped more significantly for 2-ringed to

4-ringed PAHs with relatively low molecular weights in rhizosphere layers in proximity to root surface of hyssop versus clover.

Table 2-4　Rhizosphere effect (*R*, %) on the PAH distribution in soil

PAHs	Clover			Hyssop		
	0~3mm	3~6mm	6~9mm	0~3mm	3~6mm	6~9mm
2-ringed PAHs	36.4	29.8	29.1	37.5	30.0	6.8
3-ringed PAHs	18.4	16.5	7.9	17.4	12.7	9.1
4-ringed PAHs	38.3	23.3	19.1	24.0	8.3	2.3
5-ringed PAHs	46.3	41.5	31.7	48.8	43.9	37.8
6-ringed PAHs				50.0	25.0	12.5

　　Note that the rhizosphere effects were higher for 2-ringed, 5- ringed, and 6-ringed PAHs, and lowest for 3-ringed PAHs in the three layers of rhizosphere soils. Smaller PAHs (2-ringed) were directly biodegraded by the enriched microorganisms in the rhizosphere, and larger PAHs (5-ringed and 6-ringed) could be co-metabolized by microbes in the presence of root exudates (Rentz et al., 2005). In contrast, the rhizosphere had a relatively weaker influence on the distribution of 3-ringed PAHs in soils. In this investigation, phenanthrene was the most abundant 3-ringed PAHs, with a concentration of 25.53 mg/kg in bulk soil. Phenanthrene is a more readily biodegradable PAH, even in bulk soils without root exudates (Dean-Ross et al., 2001; Corgié et al., 2003), which could have reduced the *R* value of phenanthrene in this investigation. However, the detailed mechanisms involved remain unclear.

　　Grasses are known to exude large amounts of soluble organic substrates, reaching 500 mg/g root dry weight after 4 weeks (Lynch and Whipps, 1990). This may represent 65% of the net CO_2 assimilated by grass (ryegrass), which diffused into the rhizosphere over 1 month (Figure 2-6) (Meharg and Killham, 1990). However, the rhizosphere effects differed between the two plants, and *R* values decreased more quickly from the hyssop rhizoplane to loosely adhering rhizosphere than observed in clover (Figure 2-6). Although root exudation was characterized within the last century, the methodology to identify individual exudate components was developed much later (Grayston et al., 1997). Major exudates include sugars, amino acids, and organic acids (Gao et al., 2011). The observed differences in *R* values between clover and hyssop could be related to the quantity and quality of root exudates, and their diffusion processes in rhizosphere soils.

The reduced concentrations of PAHs in the rhizosphere versus bulk soil were indicative of significant degradation of these chemicals in rhizospheric areas. The concentrations of PAHs in layers of rhizosphere soils could be calculated as follows

$$C_{rhizo} = C_{initial} - C_{biodegrad} - C_{chemdegrad} - C_{plant} - C_{abiotic}$$

where $C_{initial}$ is the initial concentration of PAHs in soil, which is equal for all layers of rhizosphere soils; $C_{biodegrad}$, $C_{chemdegrad}$, and C_{plant} are the reduction in concentrations of PAHs due to microbial biodegradation, chemical degradation, and plant uptake, respectively, in rhizosphere soils; and $C_{abiotic}$ is the decrease in PAH concentrations due to abiotic processes, excluding chemical degradation in the rhizosphere. Plant uptake contributes negligibly to the dissipation of PAHs in soils (Gao and Zhu, 2005). Abiotic dissipation (excluding chemical degradation) accounts for a minor loss of PAHs in soil, and few differences exist between the rhizosphere and bulk soils (Reilley et al., 1996). Thus, $C_{abiotic}$ is negligible in Equation 2. Thus,

$$C_{rhizo} \approx C_{initial} - C_{biodegrad} - C_{chemdegrad}$$

Clearly, differences in PAH concentrations in layers of rhizosphere soil depend on the microbial ($C_{biodegrad}$) and chemical ($C_{chemdegrad}$) degradation in these areas.

The increased gradient of PAH concentrations in layers of rhizosphere soil was primarily ascribed to the enhanced microbial biodegradation in proximity to the root surface, which was discussed in "2.1.3". The decreasing bacterial gradient reduced biodegradation and increased the gradient concentration of PAHs in layers of rhizosphere soils with increasing distance from roots. The relationships among the gradients of root exudates, microbial activity, and PAH distribution in the rhizosphere are depicted in Figure 2-7.

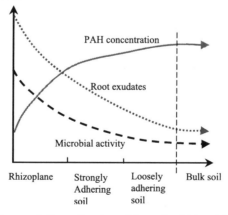

Figure 2-7 Diagram of the correlations of root exudates, microbial activity, and PAH concentration in the rhizosphere

In addition, excluding root exudates, plant roots are known to produce enzymes such as peroxidases (PODs) and polyphenol oxidases (PPOs), which can induce the chemical oxidation of aromatic rings (Liste and Alexander, 2000). Utilizing an *in vitro* study, we demonstrated that POD and PPO enzymes could effectively decompose naphthalene (NAP), phenanthrene (PHE), and anthracene (ANT) in solution (Gao et al., 2012). Similarly, Moen et al. (1994) reported that PHE could be metabolized by manganese peroxidase (MnP) in *Phanerochaete chrysosporium* in a lipid peroxidation-dependent processss. MnP from *Nematoloma frowardii* degraded ANT, PHE, pyrene, fluoranthene, and benzo[a]pyrene, leading to partial mineralization of these persistent compounds (Sarkar et al., 1997; Hofrichter et al., 1998). However, the root produced enzymes radially, which consequently diffused into the rhizoplane, as well as strongly and loosely adhering to rhizosphere soils. This partially contributed to the gradient distribution of PAHs in the rhizosphere in proximity to roots (Tables 2-2 and 2-3 and Figure 2-5).

In summary, this investigation first examined the *in situ* gradient distribution of PAHs in contaminated field rhizosphere soil in proximity to the root surface. Our results (in conjunction with our potted findings) demonstrated an increasing concentration distribution in rhizosphere soils with increasing distance from the plant root surface. Results of this work provided new information on the fate of PAHs in rhizosphere.

2.3 Rhizospheric gradient distribution of bound-PAH residues in soils

During PAH degradation in soil, bound or nonextractable residues are formed. The International Union of Pure and Applied Chemistry (IUPAC) definition reserves the term bound residues for the parent compound and its metabolites that cannot be extracted from soil using organic solvents (Roberts, 1984; Nowak et al., 2011). Similarly, a working group on bound pesticide residues in soil (Führ et al., 1998) defined bound residues as representing "compounds in a soil, plant, or animal, which persist in the matrix in the form of a parent substance or its metabolite(s) after extractions. The extraction method must not substantially change the compounds themselves or the structure of the matrix." Bound residues have a direct effect on the long-term partitioning behavior, bioavailability, and toxicity of organic contaminants in soil (Suflita and Bollag, 1981; Führ et al., 1998; Pignatello, 1998).

The distribution of organic contaminants including PAHs in the rhizosphere has been reported (Gao et al., 2011; Ling et al., 2013). However, the documented gradient distributions in rhizosphere soils overwhelmingly focus on the total concentrations of PAHs as well as other organic compounds, and few studies have hitherto addressed the distribution gradient of their bound residues in the rhizosphere.

Here, we investigated the rhizospheric gradient distribution of bound-PAH residues (ref. to parent compounds) in soils (Gao et al., 2013). Soils were collected from Nanjing, China. The soil type is TypicPaleudalf, a typical zonal soil in East China, with a pH of 6.02, 14.3 g/kg soil organic carbon content, 24.7% clay, 13.4% sand, and 61.9% silt. Soils were sieved, spiked with a solution of phenanthrene and pyrene in acetone, and aged with 20% of the soil water-holding capacity for 30 days at 22 ℃ in the dark. The final concentrations of phenanthrene and pyrene in soils were 11.90 mg/kg and 10.28 mg/kg, respectively. Then soils were poted and vegetated with Ryegrass (*Lolium multiflorum* Lam.). Soils were sampled at 30~75 days after sowing. Rhizosphere soils were collected according to the literature (Joner and Leyval, 2003; Gao et al., 2011; Ling et al., 2013).

2.3.1 Gradient distribution of bound-PAH residues in rhizosphere

Rhizosphere soils were divided into three fractions according to distance from the root surface: rhizoplane, strongly adhering and loosely adhering soil. The corresponding soil depths were 6~9 mm, 3~6 mm, and 0~3 mm, respectively. The concentrations of PAH-bound residues in soil increased with distance away from the root surface to the bulk soil (Figure 2-8). This trend was more significant at 30 days, 45 days and 60 days, while it weakened at 75 days for both phenanthrene and pyrene. For example, the concentrations of bound residual phenanthrene and pyrene in the loosely adhering soil were 12% and 42% higher, respectively, than those in the rhizoplane soil after 30 days. They were respectively 17% and 38% higher after 45 days and 11% and 9% higher after 60 days, while they were only 2% and 3% higher after 75 days. This suggests that the rhizosphere influenced the distribution of PAH-bound residues in soils.

The rhizosphere effect (R, in percent) of the bound residues of phenanthrene and pyrene in proximity to the root surface are shown in Figure 2-9. A larger R value indicates a more significant rhizosphere effect on the distribution of bound-PAH residues in soils (Ling et al., 2013). The R values decreased with distance away from the root surface for both PAHs and for all treatment durations. A more significant

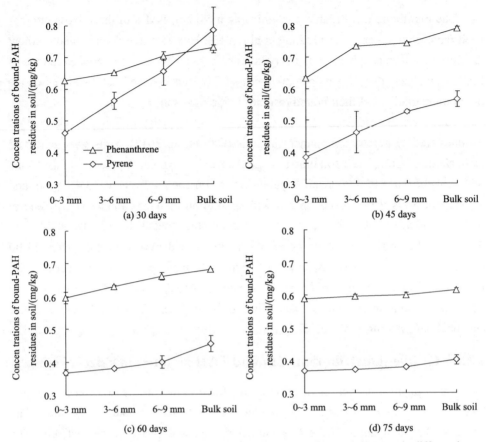

Figure 2-8　Gradient concentrations of bound residues of phenanthrene and pyrene in different layers of rhizosphere soils after 30 days (a), 45 days (b), 60 days (c), and 75 days (d). Error bars represent standard deviations (SD)

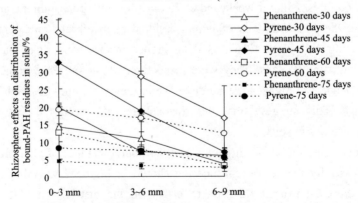

Figure 2-9　Rhizosphere effects (%) on the distribution of bound residues of phenanthrene and pyrene in soils after 30~75 days. Error bars represent standard deviations (SD)

rhizosphere effect with higher R values of bound-PAH residues was observed at 30~60 days, while R values for all three fractions of rhizosphere soils were less than 8% at 75 days. The R values of pyrene were always much larger than those of phenanthrene, suggesting a more significant rhizosphere effect on bound residual pyrene in soils.

The concentrations of bound-PAH residues in the rhizosphere as a function of time over 30~75 days are shown in Figure 2-10. The bound residual concentrations of phenanthrene and pyrene in soil generally decreased over 30~75 days for all fractions of rhizosphere soils (except for the strongly adhering soil fraction at 45 days). For example, the respective concentrations of bound residual pyrene in loosely adhering, strongly adhering and rhizoplane soil after 30 days were 73%, 51% and 25% larger than the concentrations after 75 days; and the concentrations of bound residual phenanthrene in these three layers after 30 days were 17%, 9% and 7% larger than the concentrations after 75 days, respectively. However, the bound residual concentrations of test PAHs in the rhizosphere always followed a descending order of loosely adhering, strongly adhering, and rhizoplane soil.

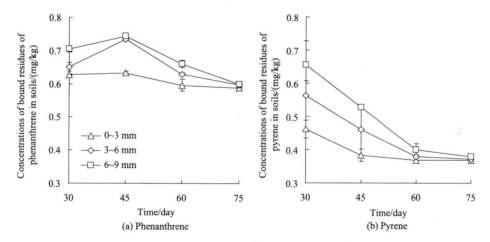

Figure 2-10 Concentrations of bound residues of phenanthrene (a) and pyrene (b) in different portions of rhizosphere soil after a plant cultivation time of 30~75 days. Error bars represent standard deviations (SD)

The dissipation ratio (D) of the bound-PAH residues in each layer of rhizosphere soils was calculated according to

$$D (\%) = (C_o - C_s)/C_o$$

where C_o is the initial concentration of the bound residue of a PAH in the soil and C_s is the residual concentration of the bound residue of a PAH in the rhizosphere after cultivation for 30~75 days. The D values as a function of time are given in Figure 2-11. After 30~75 days, 24%~40% and 0~36% of the bound residues of phenanthrene and pyrene had disappeared in the three fractions of rhizosphere soil, respectively. In addition, the D values clearly decreased in soils with increasing distance from the root surface, that is, bound-PAH residues in rhizoplane soils were more likely to dissipate than those in the loosely or strongly adhering soils.

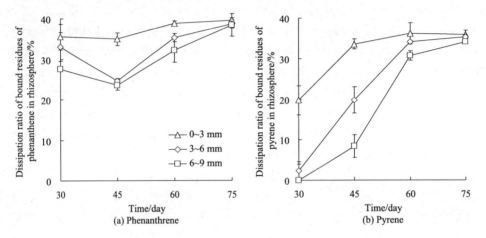

Figure 2-11 Dissipation ratio /% of bound residues of phenanthrene (a) and pyrene (b) in different portions of rhizosphere soils after a plant cultivation time of 30~75 days. Error bars represent standard deviations (SD)

Bound residues represent a minor proportion of the total concentration of PAHs in rhizosphere soils. As shown in Table 2-5, the proportions of bound residues of phenanthrene and pyrene in loosely adhering, strongly adhering, and rhizoplane soils after 30~75 days were less than 36% and 28%, respectively, while the respective proportions of the available fraction were greater than 63.6% and 72.1%. This indicates that the available fraction of test PAHs in the rhizosphere predominated. Moreover, when compared with the bound residues (Figure 2-11), the dissipation ratios of the available fractions of phenanthrene and pyrene in the three rhizosphere soil fractions were higher (Figure 2-12), suggesting the available fractions to be much more biodegradable in the rhizosphere soil.

Table 2-5 Proportions of different forms of phenanthrene and pyrene in each portion of rhizosphere soil

PAHs	Time (d)	Bound residue /%				Available fraction /%			
		0~3mm	3~6mm	6~9mm	Bulk soil	0~3mm	3~6mm	6~9mm	Bulk soil
Phenanthrene	0	—	—	—	8	—	—	—	92
	30	24	24	22	21	76	76	78	79
	45	33	36	36	35	67	64	64	65
	60	22	28	20	16	78	78	80	84
	75	27	26	26	25	73	734	74	75
Pyrene	0	—	—	—	6	—	—	—	94
	30	8	8	8	9	92	92	92	91
	45	15	12	8	7	85	88	92	93
	60	27	26	22	18	73	74	78	82
	75	28	27	26	17	72	73	74	83

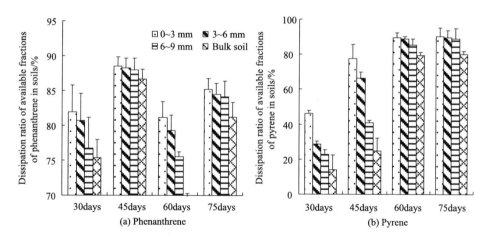

Figure 2-12 Dissipation ratios of available fractions of phenanthrene (a) and pyrene (b) in different portions of rhizosphere soils after a plant cultivation time of 30~75 days. Error bars represent standard deviations (SD)

2.3.2 Mechanism of rhizospheric gradient distribution of bound-PAH residues in soils

Studies have reported that root exudates affect the availability of recalcitrant organic pollutants, including PAHs, in the rhizosphere (Ouvrard et al., 2006; Ling et al., 2009; Kong et al., 2013). The mechanism of root exudate-influenced availability of PAHs in soil is related to a reduction in soil organic matter (SOM), the metal dissolution, and the release of dissolved organic matter from soil solids (Sun et al., 2012). The degradation of bound-PAH residues in the rhizosphere could also be related to the reduced SOM by root exudates and enhanced microbial activity in the rhizosphere soil. The SOM serves as a pool of bound residues of hydrophobic organic compounds in soil (Chiou et al., 1998). Metallic cations in soil form complexes with functional groups of soil organic molecules, and this association leads to the formation of "bridges" between minerals and the SOM (Saison et al., 2004). Root exudates, particularly organic acids, can bind metal cations. When released to soils, root exudates may dissolve the metal cations (White et al., 2003) and break the bridges between the solid surface of soil particles and the organic matter, resulting in the release of SOM into solution. Our previous findings provided evidence that the root exudates (e.g., low-molecular-weight organic acids) can significantly increase the concentrations of metal cations in soil solution (Sun et al., 2012), which supports this theory. The dissolution of some soil minerals by root-secreted organic acids has also been reported (Drever and Stillings, 1997), which theoretically reduces SOM. As a consequence, the release of SOM into solution results in the associated bound-PAH residues in soils becoming available and degradable.

Because root exudates are a convenient source of carbon and energy, they may represent an important resource for microbes and consequently alter the number of microbes and species residing in the rhizosphere (Yoshitomi and Shann, 2001; Pritchina et al., 2011). Enhanced microbial activity in the rhizosphere due to root exudates can promote the release of bound-PAH residues in soils by digestive extraction and excretion, SOM metabolism, and bioturbation (Northcott and Jones, 2000), and can enhance the biodegradation of the released PAHs from their bound residues in the rhizosphere.

Root exudates diffuse into rhizosphere soils where they gradually disappear due to radial dilution and microbial consumption, resulting in decreasing concentration gradients of root exudates in the rhizosphere as a function of proximity to the roots

(Joner and Leyval, 2003; Gao et al., 2011). A bacterial gradient has been observed with higher numbers of heterotrophs and PAH-degrading bacteria closest to the roots (Joner et al., 2001; Corgié et al., 2003; Corgié et al., 2004). Because root exudates can release bound-PAH residues and enhance microbial activity in soil, increased concentration gradients and weakened rhizosphere effects (R) of bound-PAH residues in the three portions of rhizosphere soil in proximity to a root surface was thus observed in this investigation.

Chapter 3 Partition of PAHs among soil, water and plant root

Sorption of nonionic organic chemicals (NOCs) including PAHs to soil is an important process that has a major influence on their transport, bioavailability, and fate in natural environments (Walter and Weber, 2002; Krauss and Wilcke, 2005; Ling et al., 2005, 2006). The sorption of NOCs in a soil-water system is believed to be governed by a mechanism where the NOC molecules partition into the soil organic matter (SOM) phase (Chiou et al., 1979; Karickhoff et al., 1979; Celis et al., 1998). It has been generally accepted that the NOC sorption increases with increasing SOM content. Soils with high organic matter serve as a huge sink of these compounds, and limit NOC availability in environments (Cox et al., 2000; Hwang and Cutright, 2004; Gao et al., 2015). In addition, the sorption is known to be dependent on the characteristics of SOM. For different soils, carbon-normalized distribution coefficients (K_{oc}) of a NOC may range over an order of magnitude, which can be attributed to the compositional differences in SOM, such as polarity and aromaticity (Murphy et al., 1990; White et al., 1997; Brion and Pelletier, 2005).

Studies have shown that NOCs in soils can enter plants primarily via a passive process, and this transport process could be treated as a series of contaminant partitions including the partition from soil to soil pore water and from soil water to plant root (Ryan et al., 1988; Chiou et al., 2001). The partition of NOCs between water and root was the first step and a determining process in the uptake of NOCs by plants. It was noted that highly lipophilic organic contaminants characterized by high 1-octanol/water partitioning coefficients (e.g., lgK_{ow} > 3) have a high tendency to be sorbed by plant root from water (Li et al., 2005; Collins et al., 2006). It has been known that plant composition, such as lipid content in particular, influences the uptake of NOCs (Chiou et al., 2001; Gao et al., 2005). In our previous studies, we observed that plant root concentration factors (RCF) of phenanthrene and pyrene were significantly correlated with the lipid contents in plant roots (Gao and Zhu, 2004; Gao et al., 2005). Other previous studies reported similar results (Simonich et al., 1994). Plant lipophilic components exist in membranes and cell walls. However, the distributions of NOCs in subcellular tissues remain unclear.

In this chapter, the partition of PAHs was elucidated between soil and water and between water and root. The effects of heavy metals and dissolved organic matter (DOM) on sorption of PAHs by soils were clarified. Results of this work provided insight into the sorption of PAHs in soils contaminated with heavy metal co-contaminants, the environmental behaviors of NOCs in soil-water system with DOM, and the partition of PAHs between plant composition and aqueous phase, which would be useful in predicting soil PAH contamination and plant root uptake of PAHs from the surrounding environment.

3.1 Sorption of PAHs by soils with heavy metal co-contaminants

Environmental contamination often leads to the appearance of mixtures of contaminants in soils. The sorption behavior of individual NOC in the multicomponent systems is generally assumed to be independent of the coexisted NOCs (Gao et al., 1998). But, at certain high concentrations the co-contaminant NOC can apparently influence the sorption of the target NOC as a result of the shifted solution properties (McGinley et al., 1993). Recently, the significantly inhibited sorption of phenanthrene by soil and sediment in the presence of co-contaminant NOCs of high-concentrations was reported in the literature (Walter and Weber, 2002).

The appearance of mixtures of NOCs and heavy metals in soils could be commonly found throughout the world. For instance, industrial enterprises, such as coking plants, have resulted in the simultaneous accumulation of PAHs and heavy metals with high concentrations in surrounding soils (Gao et al., 2003; Mattina et al., 2003; Ling and Gao, 2004; Saison et al., 2004). It has been known that NOCs mainly interact with SOM, and heavy metallic cations can also be complexed by the organic molecules in soil solids or released into soil solution (Gao et al., 2003; Zhang and Ke, 2004). Thus it can be postulated that the presence of heavy metals in soil may play an important role in the sorption of NOCs by soils. However, little information is available on the sorption of NOCs including PAHs by soils contaminated with heavy metals.

Here, the effects of heavy metals on sorption of PAHs by several soils were determined (Gao et al., 2006). The mechanisms involved were evaluated, based on the observed variation of DOM in solution and soil organic matter (SOM) in the presence of heavy metals. The distribution coefficient ($K_{ph/soc}$) of PAHs between water and the sorbed DOM on soil solids was firstly calculated, and was compared to the distribution coefficient (K_d) between water and SOM. Results of this work provided insight into sorption of PAHs in soils contaminated with heavy metal co-contaminants.

Three natural surface soils were experimented. Black soil (soil 1) was collected from Changchun (0~20 cm). Yellow-brown soil (soil 2) was from Nanjing (0~10 cm). Red soil (soil 3) was from (0~15 cm) Jinxian. The soil organic carbon contents (f_{oc}) of soil 1, soil 2 and soil 3 were 2.55 g/kg, 1.39 g/kg, and 0.46 g/kg, and their $pH_{2.5}$ values were 6.2, 5.2 and 4.4, respectively. Some soil samples were spiked with Pb, Zn and Cu, respectively, according to the traditional method given in literatures (Gao et al., 2003; Saison et al., 2004). Batch experiments were conducted to determine the sorption of phenanthrene as a representative PAH in soil-water systems (Gao et al., 2006). Tubes containing 1 g of soil in 15 mL phenanthrene solution with 0.02 mol/L KCl were shaken in the dark for 24 h at 250 r/min on a gyratory shaker to reach the equilibrium state. The solution and soil were separated for analysis. The losses of phenanthrene by photochemical decomposition, volatilization, and sorption to tubes were found to be negligible.

3.1.1 Sorption isotherms of phenanthrene by soils

Sorption isotherms of phenanthrene by three test soils are shown in Figure 3-1. As seen, the sorption isotherm could be well described using linear distribution-type model (Chiou et al., 1979) with the correlation constants (R) larger than 0.996. The sorption of NOCs in a soil-water system is believed to be governed by a mechanism

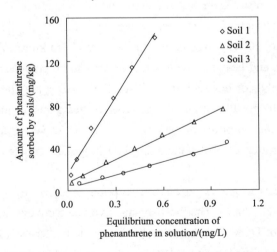

Figure 3-1 Linear isotherms for phenanthrene sorption by tested soils

where the NOC molecules partition into the SOM (Chiou et al., 1979; Karickhoff et al., 1979), and the distribution coefficient (K_d; L/kg) of solute between soil and water according to the linear sorption model is expressed as

$$K_d = Q / C_e$$

where Q denotes the amount of NOC sorbed by soil solids (mg/kg); C_e is the equilibrium concentration of NOC in aqueous phase (mg/L). In this study, the values of K_d for phenanthrene sorption by soil 1, soil 2 and soil 3 were 235 L/kg,, 70.5 L/kg, and 38.8 L/kg, respectively. Clearly, the K_d values for tested soils were positively correlated to the soil organic carbon contents, indicating the dominant influence of organic matter on phenanthrene sorption (Ling et al., 2005).

The distribution coefficient can also be expressed as

$$K_{oc} = K_d / f_{oc}$$

where K_{oc} is the organic carbon-normalized distribution coefficient (L/kg). The larger of K_{oc} means the higher capacity of per soil organic matter to sorb NOC. f_{oc} is the fractional organic carbon content of soil (g/kg). The calculated K_{oc} values for soil 1 and 3 were 9219 L/kg and 8353 L/kg, about 82% and 65% larger than the K_{oc} values for soil 2 (5064 L/kg) (Table 3-1). These results suggest that except for the total contents of SOM, the characteristics of SOM may also play a key role in NOC sorption (Nemeth-Konda et al., 2002; Ling et al., 2006).

Table 3-1 Regression data for linear sorption isotherms for phenanthrene sorption by tested soils. K_d' and K_{oc}' is the distribution coefficient and carbon-normalized distribution coefficient for phenanthrene sorption by metal-spiked soils, respectively

Soils	Soil treatment/(mg/kg)	Correlation constant (R)	K_d'/(L/kg)	K_{oc}'/(L/kg)
Soil 1 (Black soil)	Control	0.996	235 (10.6)	9219 (414)
	500 Pb [a]	0.998	252 (8.41)	9889 (330)
	500 Zn	0.962	293 (17.0)	11461 (667)
	500 Cu	0.996	252 (11.3)	9854 (443)
Soil 2 (Yellow-brown soil)	Control	0.998	70.5 (2.02)	5064 (146)
	500 Pb	0.998	78.3 (2.05)	5622 (147)
	500 Zn	0.997	79.3 (2.60)	5696 (187)
	500 Cu	0.995	76.1 (3.40)	5469 (244)
Soil 3 (Red soil)	Control	0.999	38.8 (0.97)	8353 (199)
	500 Pb	0.999	40.7 (0.87)	8764 (178)
	500 Zn	0.999	42.3 (1.21)	9105 (247)
	500 Cu	0.998	42.9 (1.45)	9238 (296)

Note: a means that soil was spiked with 500 mg Pb per kg soil. The same as follows. Data in brackets were standard errors.

The great variation of K_{oc} values of a single hydrophobic organic compound for different soils has also been reported in the literature. Such variations can be attributed to the nature and compositional differences in soil organic matter. It has been generally accepted that the nature or location of natural organic matter in soil affects the effective level of active organic matter to sorb NOCs (Chiou et al., 1986; Murphy et al., 1990; Brion and Pelletier, 2005). According to Xing and Pignatello (1997), humic substances have been described as having expanded and condensed regions, which may be analogous to rubbery and glassy states of polymers. Rubbery states of humic substances are more reactive than glassy states in the sorption of organic molecules. Moreover, some of the organic matter may be inaccessible to NOC if it associates within solid-state humic particles or clay aggregates (Spark and Swift, 2002; Ling et al., 2006). Alternatively, there would be reactive to non-reactive organic matter present in soil. The larger values of K_{oc} for soil 1 and soil 3 than soil 2 suggest that these two soils have higher proportions of reactive organic matter than soil 2.

3.1.2 Sorption of phenanthrene by heavy metal-contaminated soils

Sorption of phenanthrene by a series of heavy metal-spiked soils was also well described by the linear sorption isotherm with correlation constants (R) larger than 0.96 (Table 3-1). This indicates that partitioning into the SOM is still the dominant mechanism of phenanthrene sorption by the contaminated soils with heavy metals. The regression parameters for phenanthrene sorption by different treated soils according to the linear distribution-type model were listed in Table 3-1. The distribution coefficients (K_d') for phenanthrene sorption by the heavy metal-spiked soils were higher than the K_d values by the unspiked ones (Control). For instance, the K_d' values for soil 1 spiked with 500 mg/kg Pb, 500 mg/kg Zn, and 500 mg/kg Cu were 252 L/kg, 293 L/kg, and 252 L/kg, respectively, about 24% higher than the control. Moreover, the K_d' values tended to increase with increasing soil heavy metal contents (Figure 3-2). The organic carbon-normalized distribution coefficients (K_{oc}') for phenanthrene sorption by all heavy metal-spiked soils were also calculated. The tendency of K_{oc}' variation as a function of soil contamination with heavy metals was similar to the K_d' (Table 3-1 and Figure 3-2).

Results of this present study indicated that the presence of heavy metals in soils enhanced the sorption of phenanthrene, and the higher contents of heavy metals would generally lead to the stronger sorption of phenanthrene by soils, based on the observed K_d' and K_{oc}' values.

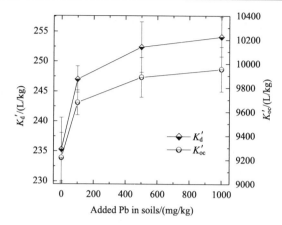

Figure 3-2 The distribution coefficients (K_d') and carbon-normalized distribution coefficients (K_{oc}') for phenanthrene sorption by Pb-spiked soil 1

3.1.3 Mechanisms of the heavy metal enhanced-sorption of phenanthrene by soils

First, sorption of phenanthrene by the deionized water-eluted soils (i.e., parts of DOM was removed off from soils) was experimented. The K_d value for phenanthrene sorption by eluted soil (taking soil 1 as an example) was 288 L/kg, about 23% larger than that by control one (235.3 L/kg). Interestingly, there was no apparent difference between sorption of phenanthrene by eluted soils with and without heavy metals (Data not shown). The results indicate that the presence of DOM in equilibrium solution inhibits phenanthrene sorption by test soils.

Concentrations of DOM in equilibrium solution (C_{doc}) for phenanthrene sorption were determined, as shown in Figure 3-3. The metallic cations have been known to complex with functional groups of organic molecules in solution, and this leads to the formation of "bridges" between soil solid surface and DOM in aqueous phase (Jones and Tiller, 1999; Gao et al., 2003; Saison et al., 2004). In this study, the C_{doc} for phenanthrene sorption by contaminated soils with co-contaminants of heavy metals were lower than those by control soils. Taking soil 1 as an example, C_{doc} for the spiked soils with 500 mg/kg Pb, 500 mg/kg Zn, and 500 mg/kg Cu were 27 mg/L, 23 mg/L, and 21 mg/L, respectively, 10%~30% lower than C_{doc} for the unspiked soil. In addition, the C_{doc} values tended to decrease with increasing concentrations of heavy metals in soils (Data not shown). The tendency of C_{doc} variation as a function of heavy metal concentrations in soils correlated negatively with phenanthrene sorption by soils, suggesting that the heavy metal-induced DOM variation in aqueous solution may account for the enhanced sorption of phenanthrene by soils in the presence of heavy metals.

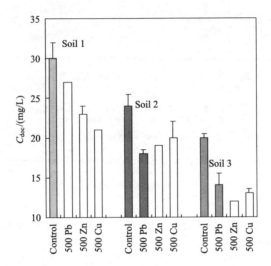

Figure 3-3　concentrations of dissolved organic carbon (C_{doc}) in solution for phenanthrene sorption by various treated soils

As mentioned above, the presence of heavy metals in soil-water system for phenanthrene sorption resulted in①the decreased DOM concentration in equilibrium solution (C_{doc}) and②increased SOM content as a consequence of DOM sorption onto soil solids (for example, through the "bridges" of metallic cations). As to the point①, it has been generally accepted that the binding of NOC with DOM in solution inhibits the sorption of NOC by soils (Celis et al., 1998; Ling et al., 2005). In our experiments, the observed decreased C_{doc} in the case of spiked soil compared to the unspiked one, in theory, would result in the reduced partitioning of phenanthrene to DOM in solution, and hence promote phenanthrene sorption in soil-water system. As to the point②, the increased soil organic carbon (SOC) content would provide new sites for phenanthrene sorption. As a result, "cumulative sorption" (Ling et al., 2006) would enhance the soil capacity of phenanthrene sorption. However, it is unknown which of these two mechanisms dominates the enhanced sorption of phenanthrene by the heavy metals.

The increased SOM contents (Q_{doc}) in the case of metal-spiked soils compared to the unspiked controls were calculated and listed in Table 3-2. Q_{doc} could be obtained

$$Q_{doc} = (C_{doc}^{\circ} - C_{doc}) \times V$$

where C_{doc}^{o} and C_{doc} are the DOM concentrations in equilibrium solution for phenanthrene sorption by control and heavy metal-spiked soils, and their values are shown in Figure 3-3; V is the volume of solution for phenanthrene sorption in soil-water systems. Thus, for all treated soils, the calculated Q_{doc} values were no more than 135 μg/g and less than 2.3% of the f_{oc} of control soils. Assuming that the increased SOM in the case of spiked soils had the same capacity of taking up phenanthrene as the original SOM in soils, and the variation of DOM in solution had little influence on phenanthrene sorption, the distribution coefficient (K'_d) for phenanthrene sorption by heavy metal-spiked soils could be simulated according to the following equation

$$K'_d = \frac{(f_{oc} + Q_{doc}) \times K_d}{f_{oc}}$$

$$\Delta K'_d = K'_d - K_d = \frac{Q_{doc} \times K_d}{f_{oc}}$$

The simulated K'_d and the calculated ΔK_d values were shown in Figure 3-4 and Table 3-2. Obviously, the predicted values of $K_d{'}$ and ΔK_d were always lower than the measured ones. This indicates that the increased SOM in the case of metal-spiked soils compared to the unspiked ones may have stronger capacity of taking up phenanthrene, and the DOM variation in solution may also contribute to the enhanced sorption of tested PAHs.

Table 3-2 Some parameters for phenanthrene sorption by tested soils

Soils	Soil treatment	Q_{doc} /(μg/g)	Q_{doc}/f_{doc} /%	$\Delta K_{d\text{-}d}/\Delta K_{d\text{-}s}$	$a \times K_{ph/doc} \times \Delta C_{doc}$	$K_{ph/soc}$ /(L/kg)
Soil 1 (Black soil)	500 Pb	45	0.176	41.2	0.020	335043
	500 Zn	105	0.412	59.0	0.047	518759
	500 Cu	135	0.529	13.0	0.060	10542
Soil 2 (Yellow-brown soil)	500 Pb	90	0.647	17.1	0.042	60249
	500 Zn	75	0.540	23.2	0.035	96875
	500 Cu	60	0.432	18.6	0.028	70690
Soil 3 (Red soil)	500 Pb	90	1.957	2.54	0.043	2212
	500 Zn	120	2.609	3.48	0.057	10508
	500 Cu	105	2.283	4.69	0.050	21903

Note: Q_{doc} is the increased SOM contents in the case of metal-spiked soils compared to the unspiked controls; f_{doc} is the fractional organic carbon content of soil g/kg; $\Delta K_{d\text{-}d}$ and $\Delta K_{d\text{-}s}$ are the measured and simulated ΔK_d values; a is a coefficient; $K_{ph/doc}$ is the association coefficient of phenanthrene with DOM in water; ΔC_{doc} is the difference in DOM concentration between the spiked and unspiked soils; $K_{ph/soc}$ is the distribution coefficient of phenanthrene between water and the sorbed DOM on soil solids.

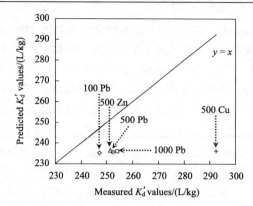

Figure 3-4 Predicted and measured values of the distribution coefficients (K_d') for phenanthrene sorption by heavy metal-spiked soils (soil 1 as an example)

Thus we introduce a modified equation provided by Celis et al. (1998) to further evaluate the mechanisms involved in promoted sorption of phenanthrene in the case of heavy metal-spiked soils.

$$K_d' = \frac{K_d + a \times K_{ph/soc} \times Q_{doc}}{1 + a \times K_{ph/doc} \times \Delta C_{doc}}$$

$$a = \frac{1}{1 + K_{ph/doc} \times C_{doc}^o}$$

where $K_{ph/soc}$ is the distribution coefficient of phenanthrene between water and the sorbed DOM on soil solids; $K_{ph/doc}$ is the association coefficient of phenanthrene with DOM in water. The $K_{ph/doc}$ value was reported to be 8317.6 L/kg (Mott, 2002); The respective C_{doc}^o values for soil 1, soil 2 and soil 3 were 30 mg/L, 24 mg/L, and 20 mg/L, respectively (Figure 3-3). Then the calculated a values for tested soils were 0.80, 0.83, and 0.86, respectively; ΔC_{doc} is the difference in DOM concentration between the spiked and unspiked soils.

In equations, the $a \times K_{ph/doc} \times \Delta C_{doc}$ values represent the contribution of DOM variation in solution to the enhanced sorption of phenanthrene by heavy metal-spiked versus unspiked soils. If $a \times K_{ph/doc} \times \Delta C_{doc} \ll 1$, the binding of phenanthrene with DOM in solution would play a negligible role in the promotion of phenanthrene sorption in the presence of heavy metals (Celis et al., 1998; Ling et al., 2006). As seen from Table 3-2, the calculated $a \times K_{ph/doc} \times \Delta C_{doc}$ values were all far smaller than 1, indicating that the DOM variation in solution was a minor mechanism of the enhanced sorption in the case of metal-spiked soils.

On the other hand, the $K_{ph/soc}$ values were obtained according to the following equation derived from above equations

$$K_{ph/soc} = \frac{(K_d' - K_d) + aK_d'K_{ph/doc}\Delta C_{doc}}{aQ_{doc}} = \frac{\Delta K_d + aK_d'K_{ph/doc}\Delta C_{doc}}{aQ_{doc}}$$

To our knowledge, there is still no direct method to determine the $K_{ph/soc}$ values heretofore, so the calculated values have to be used. As seen from Table 3-2, the calculated $K_{ph/soc}$ values for soil 1, soil 2, and soil 3 were $0.1\sim5.2\times10^5$ L/kg, $6.0\sim9.7\times10^4$ L/kg, and $0.2\sim2.2\times10^4$ L/kg, respectively, generally 2~3 magnitudes larger than their corresponding K_d values (Table 3-1). As stated previously, the natural organic matter in soil may have different capacity of NOC sorption in soil-water systems, and there would be reactive to nonreactive organic matter present in soil (Chiou et al., 1986; Murphy et al., 1990; Spark and Swift, 2002; Brion and Pelletier, 2005). The large values of $K_{ph/soc}$ compared to K_d suggest that the sorbed DOM on soil solids in the presence of heavy metals are much more reactive and do have far stronger capacity of phenanthrene sorption than the inherent SOM, which may be the dominant mechanism of the enhanced sorption of phenanthrene by heavy metals.

In summary, the enhanced sorption of phenanthrene in the case of heavy metal-spiked soils could be primarily attributed to the decreased DOM in solution and increased SOM as a consequence of DOM sorption onto soil solids. However, the decreased DOM in solution contributed little to the enhancement of phenanthrene sorption by heavy metal-spiked versus unspiked soils. On the contrary, the sorbed DOM on soil solids was found to be much more reactive and had far stronger capacity of phenanthrene uptake than the inherent SOM. Although there is still lack of information on the impacts of ageing process of heavy metals in soils on NOC sorption, the preliminary study would be helpful to the further assessment and remediation of soils polluted with co-contaminants of NOCs and heavy metals.

3.2 Dissolved organic matter(DOM)influences the partition of PAHs between soil and water

In recent years, growing attention has been given to the influence of dissolved organic matter (DOM) on sorption of NOCs by soil. Some studies revealed that the presence of DOM promoted sorption of NOCs such as nonionic pesticides (Murphy and Zachara, 1995; Totsche et al., 1997). By contrast, in other cases the sorption of NOCs was obviously inhibited in the presence of DOM (Celis et al., 1998; Gao et al.,

1998; Nelson et al., 1998). Cox et al. (2000) reported that DOM reduced sorption of herbicide due to DOM-herbicide interactions and/or competition for sorption sites on soil particles. Spark and Swift (2002) observed that the presence of DOM had little effect on sorption of atrazine, isoproturon and paraquat by soils. Similarly, Seol and Lee (2000) found that DOM (up to 150 mg DOC/L) did not significantly suppress the sorption of either atrazine or prometryne by soil solids. Clearly, results on sorption of NOCs by soils in the presence of DOM are not identical. But the mechanisms involved still need to be well defined.

The influence of DOM on sorption of NOCs by soil up-to-date mostly focused on nonionic pesticides. Only very limited information is available on the influence of DOM on sorption of PAHs by soils. In addition, the experimented DOMs in literatures are generally exotic, particularly deriving from organic composts, sediments, sewage sludges, and water from waste disposal sites (Celis et al., 1998; Gao et al., 1998). In fact, total soil organic matter includes both the soluble and insoluble fractions of organic matter, although the proportion of soluble fraction in a soil is relatively very small (Spark and Swift, 2002). However, to our best knowledge, the impacts of this soluble organic matter, i.e., the soil inherent DOM, on distribution of NOCs in three-phase system including soil, water and DOM are heretofore still under elucidation.

Here, we seek to determine the effects of exotic and inherent DOM on sorption of PAHs by a series of soils differing in organic matter contents (Gao et al., 2007). The mechanisms involved were evaluated, based on the observed distribution of DOM between soil and water. Results of this work would provide insight into the environmental behaviors of NOCs in three-phase system with DOM.

Six natural surface soil samples were collected and experimented in this study. A summary of the characteristics of these soil samples is shown in Table 3-3. To evaluate the impact of soil inherent DOM on sorption of PAH by soils, the deionized water-eluted soil samples were prepared according to Gao et al. (2007). The total organic carbon in effluent, i.e., the inherent DOM eluted off from soils (T_{doc}), were detected and listed in Table 3-4. The soil organic carbon contents of these eluted soils (f_{oc}^*) were calculated and also shown in this Table 3-4. The exotic DOM was extracted from the straw waste (Gao et al., 2007). The content of DOM in the filtered solution was 212 mg DOC/L. Batch experiments were conducted to determine the sorption of phenanthrene as a representative PAH by treated soils including the control and eluted ones. Tubes containing 1 g of soil in 15 mL of 0.02 mol/L KCl solutions with 0.05% NaN_3 and a given exotic DOM and phenanthrene concentrations were shaken in the dark for 24 h at

250 r/min on a gyratory shaker to reach the equilibrium state. The solution and soil were separated for analysis. The losses of phenanthrene by photochemical decomposition, volatilization, and sorption to tubes were found to be negligible.

Table 3-3 Some characteristics of the experimented soils

Soil No.	Location (City, province)	pH values	f_{oc}/(g/kg)
Soil 1	Jinxian, Jiangxi	4.43	4.90
Soil 2	Nanjing, Jiangsu	5.19	13.9
Soil 3	Hangzhou, Zhejiang	6.00	13.5
Soil 4	Changchun, Jilin	6.18	19.1
Soil 5	Shengzhou, Zhejiang	6.32	22.9
Soil 6	Haerbin, Heilongjiang	5.50	41.7

Table 3-4 Regression data for phenanthrene sorption by deionized water-eluted and control soils using Linear-type sorption isotherms

Soil No.	Control soil			Eluted soil				
	R^2	K_d /(L/kg)	K_{oc} /(L/kg)	T_{doc} /(g/kg)	f_{oc}^* /(g/kg)	R^2	K_d^* /(L/kg)	K_{oc}^* /(L/kg)
Soil 1	0.9863	46.05	9398	0.390	4.51	0.9831	47.49#	10530#
Soil 2	0.9959	70.49	5071	0.555	13.4	0.9696	75.48#	5656#
Soil 3	0.9987	112.9	8365	0.660	12.8	0.9925	119.6	9311
Soil 4	0.9987	122.7	6425	0.495	18.6	0.9887	136.2	7319
Soil 5	0.9878	185.5	8101	0.795	22.1	0.9979	211.1	9552
Soil 6	0.9279	310.5	7456	0.840	40.8	0.9698	377.1	9241

Note: T_{doc} is the the amounts of inherent DOM eluted off from soils; f_{oc}^* is the calculated soil organic matter contents in eluted soils; R^2 is the simulated correlation constants for phenanthrene sorption by Linear equation; # means that K_d^* or K_{oc}^* value was not significantly different from its corresponding K_d or K_{oc} values for the same soil ($p<0.05$).

3.2.1 Effect of inherent DOM on phenanthrene sorption by soils

First, the sorption isotherms of phenanthrene by test six soils were obtained. over the range of concentrations the sorption isotherms of phenanthrene by six tested soils could be well described using Linear distribution-type model (Table 3-4). The distribution coefficient (K_d; L/kg) of solute between soil and water according to the Linear sorption model is expressed as

$$K_d = Q / C_e$$

Where, Q denotes the amount of NOC sorbed by soil solids (mg/kg); C_e is the equilibrium concentration of NOC in aqueous phase (mg/L). In this study, the values of K_d for phenanthrene sorption by tested soils were 46.05~310.5 L/kg, and were in the order of soil 6>soil 5> soil 4> soil 3> soil 2> soil 1 (Table 3-4). This order significantly correlated to the order of relative contents of soil organic carbon (SOC) with a statistical R^2 value for fit of 0.96 (Figure 3-5), indicating that the soil organic matter (SOM) dominates phenanthrene sorption by soils (McGinley et al., 1993). The values of the corresponding carbon-normalized distribution constant (K_{oc}; L/kg) were also given in Table 3-4. K_{oc}, i.e., the ratio of the K_d value to the SOC content, can be expressed as

$$K_{oc} = K_d / f_{oc}$$

where f_{oc} is the fractional organic carbon content of soil (%). Clearly, the calculated K_{oc} values for six soils were dramatically closer than their corresponding K_d values. The averaged K_{oc} values were 7469±1929 L/kg (representing 26% variation). However, soil 1 (9398 L/kg) gave the highest K_{oc} value, which was still 85.3% larger than that for soil 2 (5071 L/kg) with the lowest K_{oc}. This suggests that except for SOM contents, the nature of SOM may also play a key role in NOC sorption, as discussed previously.

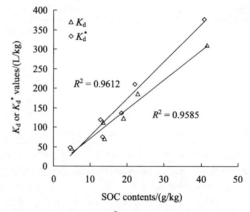

Figure 3-5   Correlation between K_d or K_d^* values and soil organic carbon (SOC) contents

The deionized water-eluted soil samples, in which the inherent DOM was removed from soils, were experimented to evaluate the influence of soil inherent DOM on phenanthrene sorption by soil. Over the range of tested concentrations, the sorption isotherms of phenanthrene by eluted soils were also fit well to Linear distribution-type model (R^2>0.97). The apparent distribution constant (K_d^*) of phenanthrene simulated from Linear equation were 47.49~377.1 L/kg, as shown in Table 3-4. Although some of the soil inherent DOM were eluted off from the soil solids, and the SOC contents of eluted soils were relatively a little (2.01%~7.96%) lower than those of the corresponding control soils (Table 3-3, 3-4), the K_d^* values were still significantly correlated to the SOC contents (Figure 3-5). This indicates that partitioning into SOM is still the dominant mechanism of phenanthrene sorption by these soils, and the SOM serves as the primary sorbent for phenanthrene sorption.

It was notable that the decrease of SOC contents due to the elution of soil inherent DOM resulted in the enhanced sorption of phenanthrene by test soils. Table 3-4 shows that the K_d^* values were 3.13%~21.5% larger than their corresponding K_d values for the same soil. This means that the presence of the soil inherent DOM impedes phenanthrene sorption by soil solids. Similarly, the calculated apparent carbon-normalized distribution constants (K_{oc}^*) for phenanthrene sorption by eluted soils were accordingly bigger than the K_{oc} values by control soils, indicating that the SOM would be strong sorptive in phenanthrene sorption when soil inherent DOM was eluted off. These results suggest that including the total SOM content, the nature of the inherent DOM in soil also plays an important role in NOC sorption, although the soils may have very small proportions of inherent DOM.

A further investigation reveals that the enhanced sorption of phenanthrene by the deionized water-eluted soils versus control soils was obviously in positive correlation with their SOC contents (f_{oc}), as shown in Figure 3-6. For instance, K_d^* value for soil 6 with the highest SOC content was 66.6 L/kg and 21.5% larger than its K_d value. By contrast, the K_d^* values for soil 1 and soil 2 were only equal to or a little larger than their respective K_d values (Figure 3-6).

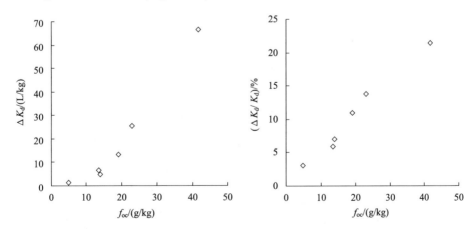

Figure 3-6 Correlations of the enhanced sorption of phenanthrene with soil organic carbon contents.
$\Delta K_d = K_d^* - K_d$

NOC sorption between soil and water is dependent on its distribution between three phases: the aqueous solution, DOM, and soil solids (Gao et al., 1998). The association of DOM with NOCs such as nonionic pesticides in solution phase is well documented, and has been proposed as an important process that results in solubility enhancement and reduced the sorption of these compounds by soil sorbents (McGinley et al., 1993; Totsche et al., 1997; Celis et al., 1998; Ling et al., 2005). There is also much evidence of the interaction of DOM with PAHs in aqueous solution. It has been suggested that the interactions of PAHs with DOM are more pronounced with its hydrophobic fractions, and the affinity of DOM to PAHs was controlled by the molecular properties of PAHs (Raber and Kögel-Knabner, 1997; Totsche et al., 1997). The reported partition coefficients (K_{doc}) of benzo(e)pyrene and benzo(k)fluoranthene with DOM in solution were 19 953~79 432 L/kg (Raber and Kögel-Knabner, 1997), and K_{doc} of phenanthrene with DOM was 8317.6 L/kg (Mott, 2002). Clearly, K_{doc} values of PAHs were generally much larger than their distribution constants (K_d or K_d^*) between soil solids and water (Table 3-4). This indicates that the strong binding affinity of PAHs with DOM in solution would impede the sorption of PAHs by soils. The higher DOM concentration in solution leads to the stronger inhibition of PAH sorption by soil solids. In this work, the enhanced sorption of phenanthrene by eluted versus control soils may also primarily be the reason that DOM concentrations (C_{doc}) in solution for the former were far lower than C_{doc} for the latter (Figure 3-7).

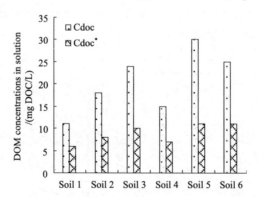

Figure 3-7 DOM concentrations in aqueous solution for phenanthrene sorption by deionized water-eluted soils (C_{doc}^*) and control soils (C_{doc})

In addition, the modified surface characteristics of solids due to the bond of inherent DOM with soil solids such as soil insoluble organic matter may also account for the enhanced sorption of phenanthrene by eluted soils. For instance, binding of inherent DOM to soil solids could take place through hydrophobic regions of DOM with hydrophilic and ionizable groups oriented to the aqueous solution, which may make the soil/water interface more hydrophilic resulting in preferential sorption of water molecules instead of PAHs (Celis et al., 1998; Gao et al., 2006; Ling et al., 2006). As such, the presence of the inherent DOM in control soils would make the surface of soil solid more hydrophilic and impede phenanthrene sorption. As to the positive correlation of the enhancement of phenanthrene sorption by eluted versus control soils with SOC contents, the involved mechanisms still needs to be further investigated.

3.2.2 Effect of exotic DOM on phenanthrene sorption by soils

The sorption of phenanthrene by tested soils was determined in the presence of exotic DOM at added concentrations ≤106 mg DOC/L. In all cases, sorption could be well described by the Linear isotherm. In the presence of exotic DOM, the apparent distribution coefficient, K_d^*, for phenanthrene sorption by soils, taking soil 5 as an example, increased first and decreased thereafter with the increase of the added DOM concentrations (0~106 mg DOC/L) (Figure 3-8). That is, the presence of exotic DOM at a low concentration (≤ 28 mg DOC/L) promoted the sorption of tested PAH. In contrast, higher exotic DOM concentrations (≥52 mg DOC/L) added in soil-water system clearly impeded the distribution of phenanthrene into soil solids. For instance, the maximum K_d^* value of 221.6 L/kg at DOM concentration of 28 mg DOC/L was 16.3% higher than its corresponding control K_d value (185.5 L/kg). Whereas the tested minimum K_d^* value at DOM concentration of 106 mg DOC/L was 15.6% lower than the control one.

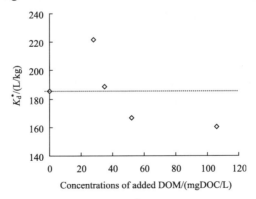

Figure 3-8　The apparent distribution constant (K_d^*) for phenanthrene sorption by soil 5 as function of the added exotic DOM concentrations

The enhanced sorption of phenanthrene by other soils with present exotic DOM of low concentrations was also observed in this work. As shown in Figure 3-9, the respective K_d^* values for phenanthrene sorption by soil 2, soil 3, and soil 4 in the presence of 28 mg DOC/L DOM were 104.6 L/kg, 129.5 L/kg, and 145.0 L/kg, which were 14.7%~48.4% larger than their corresponding K_d values. By contrast, significant impediment of phenanthrene sorption was observed when high exotic DOM present. For example, the K_d^* values for soil 4 and soil 6 in the presence of 106 mg DOC/L were 11.9 L/kg and 50.3 L/kg lower than their control K_d values, respectively.

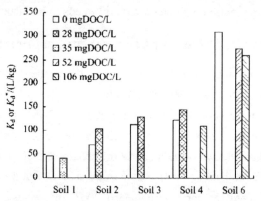

Figure 3-9 The K_d^* and K_d values for phenanthrene sorption by soils at various exotic DOM concentrations

Results above were somewhat different from previous findings on sorption of NOCs by soils in the presence of exotic DOM. Celis et al. (1998) reported that the added DOM from sewage sludge decreased atrazine sorption by soils, and similar results were reported in literatures (Gao et al., 1998; Nelson et al., 1998). By contrast, the enhanced sorption of NOCs was also well documented (Totsche et al., 1997). Others observed that the presence of DOM had little effects on NOC sorption (Seol and Lee, 2000; Spark, and Swift, 2002). In this work, we found that the impacts of exotic DOM on phenanthrene sorption by tested soils actually were DOM concentration-dependant (Figures 3-8 and 3-9).

The influence of exotic DOM on NOC sorption may be ascribed to their interaction in solution, competition for sorption sites, co-sorption, and cumulative sorption (McGinley et al., 1993; Totsche et al., 1997; Celis et al., 1998; Ling et al., 2005). As discussed previously, the association of DOM with NOC in solution may decrease the NOC sorption by soil solids. In addition, the competition of DOM with relatively polar NOC such as some pesticides for sorption sites also tended to reduce its sorption by soil (Gao et al., 1998; Celis et al., 1998; Spark and Swift, 2002). However, as a highly hydrophobic organic compound with a $\lg K_{ow}$ of 4.46 (Yaws, 1999), phenanthrene sorption was closely related to SOC content irrespective of the presence of DOM, i.e., organic matter dominates phenanthrene sorption by soil. While the clay mineral surface generally contributes to the sorption of DOM (McGinley et al., 1993; Ling et al., 2005). As such, competition of phenanthrene with DOM for sorption sites seems to be a minor mechanism of the impacts of exotic DOM on phenanthrene sorption. Additionally, the DOM-mediated sorption (co-sorption), i.e., the formation of a DOM-NOC complex and its sorption by soil which is suggested to increase NOC sorption, did not seem to be a major factor controlling phenanthrene sorption, since sorption in the presence of DOM could be well described by the Linear model (Ling et al., 2006).

The sorption of DOM in soil-water system was observed, and its sorption amounts (Q_{doc}) were calculated and shown in Table 3-5. The DOM sorption to soil solids may increase the bulk SOC contents, and hence provide new sorption sites. As a result, such "cumulative sorption" would increase the soil's capacity of taking up phenanthrene and promote its sorption. In this work, the added exotic DOM at low concentrations ($\leqslant 28$ mg DOC/L) was primarily sorbed to soil solids. The initial rise in K_d^* values for phenanthrene sorption by tested soils in the range of tested DOM concentrations, as showed in Figures 3-8 and 3-9, was consistent with the above suggestion.

Table 3-5 Some parameters for phenanthrene sorption by soil 5 in the presence of exotic DOM at various concentrations

Added DOM concentrations /(mg DOC/L)	K_d^* /(L/kg)	C_{doc}^* /(mg DOC/L)	Q_{doc} /(mg DOC/kg)	$K_{ph/soc}$ /(L/kg)
0	185.5	30		
28	221.6	48	150	461848
35	188.6	60	75	668813
52	166.6	69	195	180219
106	160.4	109	405	198265

Note: C_{doc}^* is the equilibrium concentrations of DOM in aqueous solution for phenanthene sorption by soil 5.

On the other hand, the increased SOM (expressed as Q_{doc}, as shown in Table 3-5) due to the DOM sorption may not have the same capacity of phenanthrene uptake as the original SOM in soils, since the enhanced SOC contents (Q_{doc}) were less than 1.77% of the soil f_{oc}, taking soil 5 as an example. We further evaluated the capacity of the sorbed DOM for phenanthrene uptake, based on the equation (Gao et al., 2006) as follows

$$K_d^* = \frac{K_d + K_{ph/soc} Q_{doc}}{1 + K_{ph/doc} \times \Delta Q_{doc}}$$

This equation could also be expressed as

$$K_{ph/soc} = \frac{K_d^*(1 + K_{ph/doc} \times \Delta C_{doc}) - K_d}{Q_{doc}}$$

where $K_{ph/soc}$ is the distribution coefficient of phenanthrene between water and the sorbed DOM on soil solids; $K_{ph/doc}$ is the association coefficient of phenanthrene with DOM in water, and its value was 8317.6 L/kg (Mott, 2002); ΔC_{doc} is the difference of DOM concentration in aqueous solution in the presence versus absence of exotic DOM. According to equation 4, the $K_{ph/soc}$ values of phenanthrene were calculated and displayed in Table 3-5. Obviously, $K_{ph/soc}$ values were about three magnitudes larger than their corresponding K_d values. As stated previously, the nature or location of organic matter in soil will affect the effective level of active organic matter to sorb NOCs (Chiou et al., 1986; Murphy and Zachara, 1995; Spark and Swift, 2002; Brion and Pelletier, 2005). The large values of $K_{ph/soc}$ compared to K_d suggest that the sorbed DOM on soil solids are much stronger sorptive and do have far stronger capacity of phenanthrene uptake than the original SOM, which may also be a major mechanism of the enhanced sorption of phenanthrene by tested soils in the presence of exotic DOM at lower concentrations.

However, in the range of tested concentrations, the K_d^* values turned to decrease after an initial increase with the extensive increase of the added exotic DOM concentrations (Figures 3-8 and 3-9). The decreased sorption may be the result of the enhancement of phenanthrene association with DOM in solution since C_{doc}^* increases straightly with increasing the added DOM concentrations (Table 3-5), as mentioned previously.

On the whole, the influence of exotic DOM on phenanthrene sorption could be approximately considered as the net effect of the "cumulative sorption" and the association of phenanthene with DOM in solution. While the co-sorption and competition between DOM and phenanthrene for sorption sites would have minor effects on phenanthrene sorption by soil solids.

3.3 Partition of polycyclic aromatic hydrocarbons between plant root and water

During the past decades, there has been a worldwide interest in understanding plant uptake of organic contaminants (Wang and Jones, 1994; Li et al., 2002; Mattina et al., 2003; Gao and Zhu, 2004; Gao and Ling, 2006). The rate and extent of plant uptake depends on physiochemical properties of organic contaminants, characteristics of soil and water, and plant species and physiology (Burken and Schnoor, 1998; Gao et al., 2005; Collins et al., 2006). Recent studies have shown that NOCs in soils can enter plants primarily via a passive process, and this transport process could be treated as a series of contaminant partitions including the partition from soil to soil pore water, from soil water to plant root, and from xylem water to shoot (Ryan et al., 1988; Chiou et al., 2001). The partition of organic chemicals between water and root was the first step and a determining process in the uptake of NOCs by plants.

Here, the partition of phenanthrene as a representative PAH between water and a series of plant roots was investigated. Partition between root cell walls and water was also determined (Gao et al., 2008). Thirteen plant species were examined for uptake of phenanthrene in this study (Table 3-6). The plants were cultivated. Roots were sampled. The fraction of root cell walls was obtained using the protocol described by Carrier et al. (2003). Roots and root cell walls were dried in an oven at 105℃ for 24 h, ground and passed through a 20-mesh sieve (Gao et al., 2008). The lipid and water contents of the roots were determined using the methods reported by Simonich and Hites (1994) and Gao and Zhu (2004). Batch experiment was conducted to determine phenanthrene partition in root-water systems (Gao et al., 2006, 2007). Twenty mL of phenanthrene solution containing 0.05% NaN_3 was mixed with 0.1 g of root or root cell wall in 30-mL glass centrifuge tubes stopped with screw caps. The tubes were shaken in the dark at 200 r/min on a gyratory shaker for 24 h to approach the equilibration. Then the solution and soil were separated for analysis.

The results of this work provided some insight into the partition of lipophilic organic contaminants between plant composition and aqueous phase, which could be useful in predicting plant root uptake of PAHs from the surrounding environment.

Table 3-6 Root compositions and phenanthrene sorption by plant roots

Plant species		f_{lip} [a]	f_{ch} [b]	$\lg K_{rt}$ [c]	$\lg K_{rt\text{-}p}$ [d]	$\lg K_{rt\text{-}lip}$ [e]
Common name	Latin name	/%	/%	/(L/kg)	/(L/kg)	/(L/kg)
Rice	*O. sativa* L.	0.51(0.07)[f]	15.7(2.33)	3.41	2.95	2.95
—	*Chrysanthemum nankingense* Hand-Mezz	1.37(0.25)	8.63(1.07)	3.28	3.60	3.60
Garlic	*Allium sativum* L.	0.65(0.05)	9.35(0.89)	2.87	3.27	3.27
Tall fescue	*Festuca arundinacea* Schreb.	0.46(0.05)	11.9(1.32)	3.12	3.03	3.03
—	*Carex siderosticta* Hance	0.51(0.07)	9.49(1.01)	3.09	3.17	3.17
—	*Zephyranthes candida* (Lindl.) Herb.	0.77(0.09)	9.23(0.79)	3.38	3.35	3.35
—	*Eleusine indica* (L.)Gaertn.	0.28(0.02)	9.72(0.66)	3.12	2.91	2.91
—	*Setaria viridis* (L.) Beauv.	0.26(0.02)	9.74(1.17)	3.13	2.88	2.87
—	*Erigeron annuus* (Linn.)Pers.	1.13(0.09)	17.9(1.43)	3.31	3.23	3.23
—	*Ophiopogon japonicus* (Thunb.) Ker-Gawl	1.02(0.13)	8.98(0.78)	3.08	3.47	3.47
—	*Euchsaena mexicana* schrad	0.70(0.07)	13.3(1.19)	3.13	3.16	3.16
Sundangrass	*Sorghum sudanense* (Piper) Stapf.	0.83(0.09)	14.2(1.79)	3.14	3.20	3.20
Ryegrass	*Lolium multiflorum* Lam.	0.43(0.06)	17.2(2.15)	3.08	2.84	2.84

Note: a,b f_{lip} and f_{ch} are the weight fractions of the lipids and the sum of carbohydrates, cellulose, and proteins in roots; c K_{rt} is distribution coefficient of phenanthrene between root and water; d,e $\lg K_{rt\text{-}p}$ and $\lg K_{rt\text{-}lip}$ are the redic- ted partition coefficients between root and water based on the whole root and lipid only, respectively; "—" means not available; f Values in brackets are standard deviations ($n = 3$).

3.3.1 Partition of phenanthrene between roots and water

Sorption of phenanthrene by the roots of *Euchsaena mexicana* schrad and other plants is presented in Figures 3-10 and 3-11. Regression was performed between the sorption amounts of phenanthrene by roots and its equilibrium concentrations in solution. As seen, sorption isotherms by roots of 13 plant species were all significantly linear ($p<0.05$), which is a characteristic of a partition-dominated process (Li et al., 2005; Gao et al., 2007). The distribution coefficient (K_{rt}, L/kg) of solute between root and water is expressed as

$$K_{rt} = Q_{eq} / C_w$$

Where Q_{eq} denotes the amount of phenanthrene sorbed by root (mg/kg). C_w is equilibrium concentration of phenanthrene in aqueous phase (mg/L). The K_{rt} values were obtained through linear regression of the sorption data, and reported in Table 1. As shown, the K_{rt} values ranged from 734 L/kg to 2564 L/kg for the roots of different plant species.

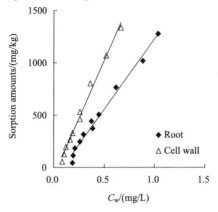

Figure 3-10   Sorption isotherms of phenanthrene by root and root cell wall of *Euchsaena mexicana* schrad. C_w was the concentrations of phenanthrene in water

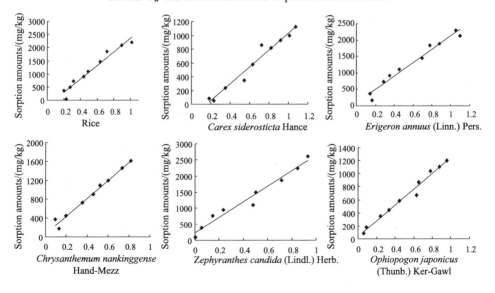

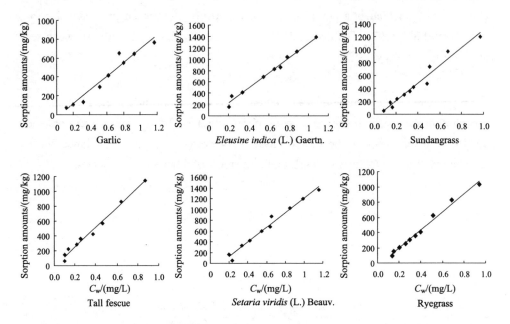

Figure 3-11 Sorption isotherms of phenanthrene by various plant roots. C_w was the concentrations of phenanthrene in water

3.3.2 Estimation of partition coefficient of phenanthrene between root and water using a composition model

Partition of organic contaminants between plant and water could be described as the distributions of the compounds in various plant organic components (Chiou et al., 2001; Li et al., 2005). The plant organic components include two major categories: lipids and carbohydrates. Therefore, K_{rt} could be described as

$$K_{rt} = f_{lip} K_{lip} + f_{ch} K_{ch} + f_w$$

Where, f_w, f_{lip}, and f_{ch} are the weight fractions of the water, lipids, and the sum of carbohydrates, cellulose, and proteins in roots that are assumed to have approximately the same partition coefficients (K_{ch}). As an approximation, the lipid/water partition coefficient (K_{lip}) is assumed to be equal to the corresponding 1-octanol/water partition coefficient (K_{ow}) (Chiou et al., 2001). Thus the above equation could be expressed as

$$K_{rt\text{-}p} = f_{lip} K_{ow} + f_{ch} K_{ch} + f_w$$

The $\lg K_{ow}$ value for phenanthrene is 4.46. According to Chiou et al. (2001), the K_{ch} values are generally small and approximated to be 3 L/kg for phenanthrene. The root compositions of the 13 plant species investigated in this study are listed in Table 3-6. The predicted partition coefficients (K_{rt-p}) using the above equation are shown in Table 3-6. The predicted K_{rt-p} (lg unit) of phenanthrene between root and water were in good agreement with the experimental results (K_{rt}, lg unit) with the differences less than 14% for all studied plants. This indicates that the partition process of phenanthrene between root and water could be accurately estimated using the root composition model (equation above).

The estimated results using above equation indicate that more than 99% of phenanthrene in roots is distributed in the lipid fraction of roots. The predominant role of lipid fractions over other root components for phenanthrene could simplify above equation to

$$K_{rt-lip} \approx f_{lip} K_{ow}$$

The estimated results (lgK_{rt-lip}) basing on the lipid partition are also listed in Table 3-6. As expected, no difference was observed between lgK_{rt-p} and lgK_{rt-lip}. Results above provided a simple but accurate method to estimate the partitions of lipophilic organic chemicals between root and water using a composition model including root lipid content and 1-octanol/water partition coefficient of the solute. The differences between the observed and estimated data for phenanthrene were <14% for 13 types of plants investigated.

3.3.3 Partition of phenanthrene between root cell walls and water

Previous studies primarily focus on the uptake and distributions of lipophilic organic contaminants in the whole plants (Wang and Jones, 1994; Li et al., 2002; Mattina et al., 2003; Gao and Zhu, 2004; Wild et al., 2005), and limited information is available to assess the distribution of these contaminants in the fractions of plant cells such as cell walls.

Cell walls were separated from roots, and used as sorbents to measure sorption for phenanthrene. Sorption isotherm by the root cell walls of *Euchsaena mexicana* schrad is also shown in Figure 3-10. The sorption isotherm was also highly linear ($p<0.05$), a characteristic of partition-dominated process (Li et al., 2005). The sorption by cell walls was apparently greater than that by the corresponding whole root. Partition coefficients of phenanthrene between cell walls and water are obtained through linear regression of sorption data, and listed in Table 3-7. The K_{rt} values for phenanthrene in cell walls were 13%~85% greater than those by the corresponding roots, indicating that cell walls have higher partition abilities for lipophilic organic chemicals.

Table 3-7   Partition coefficients of phenanthrene by roots and root cell walls

Plant species	$K_{\text{rt-cell wall}}$ /(L/kg)	$K_{\text{rt-root}}$ /(L/kg)	E/%
Tall fescue	2424 a	1320 b	83.6
Ryegrass	1346 a	1194 a	12.7
Euchsaena mexicana schrad	2209 a	1355 b	63.0
Sundangrass	1769 a	1376 b	28.6

Note: $K_{\text{rt-root}}$ and $K_{\text{rt-cell wall}}$ were the partition coefficients for phenanthrene sorption by root and root cell wall, respectively; E was the percentage of the enhancement of $K_{\text{rt-cell wall}}$ versus $K_{\text{rt-root}}$. Different letter in the same line meant significant difference between $K_{\text{rt-cell wall}}$ and $K_{\text{rt-root}}$ ($p < 0.05$).

The fraction of cell walls is the major portion of the whole plant root tissues. For instance, we observed that cell walls contributed 91% to the root tissues of ryegrass on the dry weight basis. As such, the results described above reveal that cell wall is the primary domain for the partition of highly lipophilic solutes such as PAHs relative to other plant root fractions, such as cell membrane and organelles. Plant lipid components are regarded as the predominant domains for uptake of highly lipophilic organic contaminants (Chiou et al., 2001; Gao and Zhu, 2004; Collins et al., 2006). The difference of partitioning ability for cell walls and whole root might result from their compositions. The fractionation method to separate the cell wall from the whole roots was referred to literatures (Lozano-Rodriguez et al., 1997; Carrier et al., 2003). One notes that there might be possibilities that the cell membrane was not adequately separated from the cell wall according to this method. Parts of the cell membrane might be combined into the cell walls, resulting in the higher lipid contents in cell walls and favoring the uptake of lipophilic compounds.

Chapter 4 Impact of root exudates on the sorption, desorption and availability of PAHs in soil

During plant growth, roots actively or passively release a range of organic compounds, referred to as root exudates (REs), that include carbohydrates, organic acids, amino acids, etc. (Zhang and Dong, 2008). Because these REs are a convenient source of carbon and energy, they may represent an important resource for fast-growing microbes and subsequently alter the species composition of the rhizosphere, which functions in nutrient transformation, decomposition, and the mineralization of organic substances (Petra et al., 2004).

The process of sorption-desorption of NOCs, including PAHs, is a most important process that has a major influence on their transport, bioavailability, and fate in natural environments (Li et al., 2004; Ling et al., 2006), and the sorption and slow desorption of PAHs in soils has been a major impediment to the successful rhizoremediation of contaminated soils. Recent studies have shown that root exudates may alter the availability of NOCs in the soil environment (Ouvrard et al., 2006; Ling et al., 2009). Joner et al. (2002) found that the concentrations of five-ring and six-ring PAHs in soil are enhanced in the presence of root exudates, and attributed this to the desorption of initially unextractable molecules. An increase in n-butanol extractable PAHs (which are bioavailable to organisms) was induced by root exudates (Ling et al., 2009; Liste and Alexander, 2002). However, excluding root exudate complexes, little information is available on the effects of root exudate components (RECs) on NOC availabilities in soil.

In this chapter, the PAH-influenced release of root exudates in the rhizosphere was firstly clarified. The impact of root exudates and root exudate components (RECs) on the sorption, desorption and availability of PAHs in soil were elucidated. Results of this work may be useful in assessing NOC-related risks to human health and the environment and in developing rhizoremediation strategies at sites contaminated with recalcitrant organic chemicals.

4.1 Impact of PAHs on root exudate release in rhizosphere

During plant growth, roots actively or passively release a range of organic compounds referred to as root exudates (Gao et al., 2003; Phillips et al., 2003), and the rhizosphere is defined as the volume of soil that is shared with soil bacteria and over which roots have an influence (Raynaud, 2010). The main drivers of rhizosphere formation are the development of water and solute gradients around roots, which can alter soil properties. Among the solutes, large quantities of readily available organic substrates in the form of root exudates are released into the rhizosphere. Up to 40% of net carbon fixed by plant shoots during photosynthesis is released into soils (Lynch and Whipps, 1990). The first report of root exudation was in the zoth century, but methodology for identifying individual exudate components has developed much later (Grayston et al., 1997). Major root exudates include sugars, amino acids, and organic acids (Gao et al., 2003; Phillips et al., 2003; Bais et al., 2006).

Here, the impacts of phenanthrene as a representative PAH on the root exudate release were examined using a greenhouse experiment (Gao et al., 2011). A typical zonal soil (Typic Paleudalf) with non-detectable PAHs was collected from the A (0~20 cm) horizon in Nanjing, China. This soil had a pH of 6.02, 14.3 g/kg soil organic carbon content (f_{oc}), 24.7% clay, 13.4% sand, and 61.9% silt. Phenanthrene contaminated soil was prepared according to the traditional methods documented in literatures (Gao et al., 2011). Ryegrass (*Lolium multiflorum* Lam.) was vegetated in pots with phenanthrene contaminated soils, and was sampled 40 days and 50 days after sowing. Root exudates in the rhizosphere soil were collected according to the methods of Szmigielska et al. (1996) and Xu et al. (2007).

4.1.1 Impact of PAH contamination levels on root exudation in rhizosphere

Root exudates in the rhizosphere soil were examined in this study. Soluble organic carbon (SOC) is commonly used to characterize the total amount of root exudates. Organic acids (OA) and total soluble sugar (TSS) are two important components of low-molecular-weight root exudates (Xie et al., 2008). Although many kinds of organic acid are secreted from roots into the soil environment, oxalic acid is the major acid secreted. Our previous work showed that oxalic acid was the main component of low-molecular-weight organic acids of ryegrass root exudates and constituted about 90% of total organic acids (Yang et al., 2010), as also reported by Xie et al. (2008). Hence, oxalic acid was selected as a representative organic acid and, together with SOC and TSS, was utilized to characterize the root exudates in PAH-contaminated rhizosphere soils.

The impact of PAH contamination levels on the root exudates, characterized as SOC, OA (oxalic acid as a representative), and TSS, in the rhizosphere was examined. Figure 4-1 shows the root exudate concentrations of ryegrass as a function of phenanthrene concentration (0~250 mg/kg) in the rhizosphere soil. Compared with the no-PAH control soils, the concentrations of SOC, OA and TSS in phenanthrene-spiked soil after 40 days and 50 days were higher. The root exudate concentrations increased initially and decreased thereafter with the increase of the spiked concentration of phenanthrene in soil from 0 mg/kg to 250 mg/kg, irrespective of cultivation time (40 days or 50 days), suggesting that low PAH concentrations led to the enhanced root exudate concentrations, but enhancement weakened when soil contamination increased.

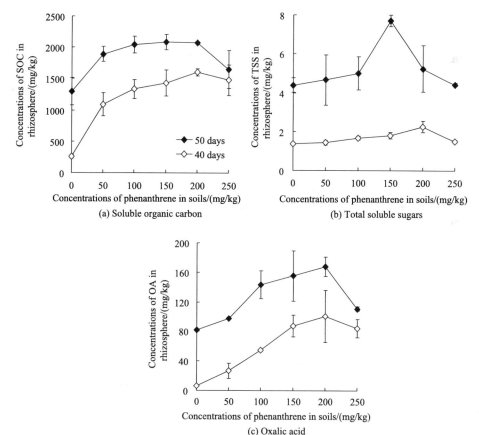

Figure 4-1 Concentrations of root exudates in phenanthrene-spiked sterilized rhizosphere soils as a function of initial phenanthrene concentrations in soils. (a) Soluble organic carbon; (b) Total soluble sugars; (c) Oxalic acid. Error bars are standard deviation (SD)

Additionally, microorganisms seem to play important roles in the richness of root exudates. As shown in Figure 4-2, the concentrations of SOC, oxalic acid, and TSS in sterilized rhizosphere soil were 123%, 58.4%, and 370% higher than those in non-sterilized soil, respectively. Root exudates provide the necessary nutrition and energy for survival and reproduction of rhizosphere microbes (Overbeek et al., 1995; Binet et al., 2000; Joner and Leyal, 2003; Gao and Zhu, 2005). In sterilized soil, microbial activity and reproduction are limited. Due to microbial activity, less root exudate was degraded in the sterilized treatment compared with the non-sterilized treatment, theoretically resulting in higher exudate concentrations in the sterilized rhizosphere.

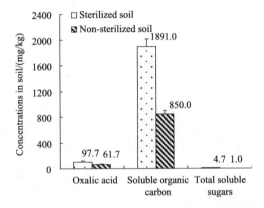

Figure 4-2 Concentrations of root exudates characterized by soluble organic carbon, total soluble sugars, and oxalic acid in phenanthrene-spiked rhizosphere soils after 50 days. The initial concentration of phenanthrene in soil was 50 mg/kg. Values of the concentrations of root exudates between sterilized and non-sterilized soil were all significantly different ($p<0.05$). Error bars are standard deviation (SD)

Root exudates respond to changes in the environment. Raynaud (2010) reported that soil properties are key determinants for development of root exudates in the rhizosphere. Lin et al. (2000) observed that the quantity of soluble sugars in root exudates of wheat increased with increasing Pb and Cd concentration and was affected distinctively by Pb and Cd co-contamination. Here, we first determined the impact of PAH contamination levels on root exudates in the rhizosphere. In our previous work with a hydroponic greenhouse experiment, we observed that the amount of root exudates from ryegrass (*Lolium multiflorum* Lam.), as characterized by soluble organic carbon, oxalic acid, and soluble carbohydrate, first increased and then decreased according to the increment of pyrene concentration in the solution (Yang et al., 2010). Here, the findings after soil cultivation further support this result that low PAH concentrations may have enhanced root exudate concentration, but enhancement weakened as soil contamination increased (Figure 4-1). PAHs may induce an increase in plant biomass production due to hormone-like effects (Fismes et al., 2002). Although the biomass of ryegrass grown in spiked soils with <150 mg/kg phenanthrene did not differ significantly from that grown in unspiked control soil, root exudate concentrations was clearly enhanced by phenanthrene (Figure 4-1). However, clear phytotoxicity symptoms were observed with a reduction of biomass for plants grown in soils with >200 mg/kg phenanthrene, and root exudate concentration was consequently reduced.

4.1.2 Distribution of root exudates in different layers of rhizosphere soil

The three layers of rhizosphere soil from the root surface (0~8 mm) were divided into rhizoplane, strongly adhering, and loosely adhering soil (Gao et al., 2011). The root exudates were characterized as SOC, OA and TSS. The gradient distribution of root exudates in the rhizosphere was elucidated in "2.1.2". On a whole, the amount of root exudates in rhizosphere soil decreases as the distance increases from root surface. Away from the root surface, root exudates diffuse radially into layers of the rhizoplane soil, strongly adhering soil, and loosely adhering soil. However, a portion of the root exudates is degraded in soils during the diffusion process, resulting in a decrease in concentrations with distance from the roots.

4.2 Impact of root exudates on PAH sorption by soils

Sorption of PAHs to soil is an important process that has a major influence on their transport, bioavailability, and fate in natural environments (Walter and Weber, 2002). The sorption of PAHs in a soil water system is believed to be governed by a mechanism where the PAH molecules partition into the soil organic matter (SOM) phase (Chiou et al., 1979; Karickhoff et al., 1979; Celis et al., 1998; Gao et al., 2006, 2007). Here, the impacts of root exudates on sorption of phenanthrene as a representative PAH by soil were investigated.

TypicPaleudalfs, a typical zonal soil previously free of PAHs, was collected from the A (0~20 cm) horizon in Nanjing, China. The soil had a pH of 6.02, 14.3% soil organic carbon content (f_{oc}), and consisted of 24.7% clay, 13.4% sand, and 61.9% silt.

A batch experiment was performed to determine phenanthrene partitioning in a soil-water system (Gao et al., 2010b, 2011). Four root exudate components (RECs) including malic acid, alanine, serine, and fructose were experimented. Solutions with 0~70.1 g/L REC (25 mL), 0.01 mol/L KNO_3 and 0.05% NaN_3 was mixed with 1.0 g of soil in a 30 mL glass centrifuge tube. A concentrated methanol stock solution was prepared with phenanthrene and added to the batch vials as required. Methanol concentrations in the aqueous solutions were always <0.2%, a level at which methanol has no measurable effect on sorption (Nkedi-Kizza et al., 1987). The tubes were shaken in the dark at 200 r/min on a gyratory shaker for 24 h to achieve equilibration. The solution and soil were separated for analysis. The losses of phenanthrene by photochemical decomposition, volatilization, and sorption to tubes were found to be negligible.

4.2.1 Root exudate component-influenced sorption of PAH by soil

It is known that the sorption of NOCs in a soil-water system is governed by NOC molecules partitioning into SOM (Chiou et al., 1979, 1998; Li et al., 2004). The distribution coefficient (K_d, L/kg) of the solute between the soil and water according to the linear sorption model is expressed as

$$K_d = Q / C_w$$

where Q denotes the amount of NOC sorbed by soil solids (mg/kg) and C_w is the equilibrium concentration of NOC in the aqueous phase (mg/L). The value of the corresponding carbon-normalized distribution constant (K_{oc}, L/kg) was calculated according to the following equation

$$K_{oc} = K_d / f_{oc}$$

where f_{oc} is the fractional organic carbon content of the soil (%). The larger the K_{oc}, the higher the capacity per unit SOM to sorb NOC.

As well as other NOC sorption by soils, the sorption isotherm of phenanthrene (over a range of concentrations) by test soil could be described using a linear distribution-type model (Figure 4-3), and the K_d value for phenanthrene sorption by test soil was 998.3 L/kg, and the calculated K_{oc} value was 71071 L/kg.

Phenanthrene sorption isotherms at different concentrations of RECs could also be described by the linear distribution-type model, indicating that partitioning into SOM remains the dominant mechanism of phenanthrene sorption by soils with RECs. The calculated values of K_d and K_{oc} for phenanthrene sorption as functions of the

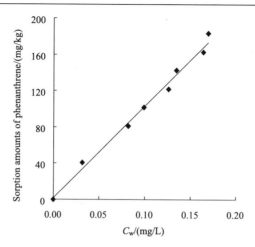

Figure 4-3 Sorption isotherm of phenanthrene by test soil. C_w is the equilibrium concentration of phenanthrene in solution

concentrations of RECs in solution are shown in Figure 4-4. As shown, the addition of RECs in solution significantly reduced the K_d values for the sorption of phenanthrene by test soil. Thus, the presence of root exudates (irrespective of the different components) in a soil-water system inhibited the distribution of phenanthrene into the soil solids. For instance, the K_d values for the sorption of phenanthrene by soil were 728.4 L/kg and 521.0 L/kg for the samples mixed with malic acid at 4.2 g/L and 67.0 g/L, which were 22% and 47% lower than the control K_d value (998.3 L/kg), respectively. Moreover, the K_d values decreased with increasing concentrations of RECs. We also calculated the K_{oc} values for phenanthrene sorption on test soil supplemented with RECs. The K_{oc} variation as a function of soil plus root exudates was similar to the K_d value (Figure 4-4). These results indicate that RECs in soils reduce the sorption potential of phenanthrene by soil, and that increasing the concentration of RECs can significantly reduce the phenanthrene sorption affinity, as determined by the observed K_d and K_{oc} values.

Additionally, different RECs inhibited phenanthrene sorption by soil to different extents. K_d and K_{oc} values with different RECs at test concentrations generally followed the order of organic acid < amino acids < fructose (Figure 4-4). Thus, organic acid had the strongest inhibitory effect on phenanthrene sorption by soil. Amino acids resulted in moderate inhibition, while fructose inhibition was low. However, all test RECs significantly inhibited phenanthrene sorption under the experimental conditions.

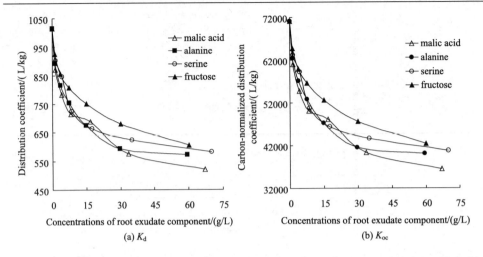

Figure 4-4 Distribution coefficient (K_d) (a) and carbon-normalized distribution coefficient (K_{oc}) (b) for phenanthrene sorption by soil as a function of increasing amount of root exudate component

4.2.2 Mechanism discussions

SOM is the predominant pool of hydrophobic organic compounds in soil (Chiou et al., 1979, 1998). It has been generally recognized that metallic cations in soil form complexes with functional groups of soil organic molecules, and this association leads to the formation of "bridges" between minerals and the SOM in soil (Saison et al., 2004). Organic acids, as the most active group of root exudates, are found in the environment as monocarboxylic, dicarboxylic, and tricarboxylic acids including compounds containing unsaturated carbon and hydroxyl groups. In theory, organic acids can bind metal cations. When added to soils, organic acids may dissolve the metal cations (White et al., 2003), breaking the "bridges" between the soil solid surface and the organic matter. Thus, organic matter complexed with soil minerals through a "bridge" of metal cations releases and becomes DOM in solution (Lu et al., 2007), resulting in decreased SOM contents in soil solids and increased DOM in solution. On the other hand, the dissolution of some soil minerals by root-secreted organic acids has been reported (Drever and Stillings, 1997), which also theoretically reduces SOM in soil and increases DOM in solution.

We measured the concentration of metal cations (C_{Metal}) in solutions of phenanthrene sorption by soils in the presence of organic acid. We observed six metals, including Al, Cu, Fe, Mg, Mn and Zn, which generally complex SOM with minerals.

As shown in Table 4-1, the C_{Metal} values of test metals significantly increased with increasing organic acid concentrations in the soil-water system, and the C_{Metal} observed values after treatment with malic acid were typically one to two orders of magnitudes larger than the controls (without root exudate addition). The above results support the concept that organic acid dissolves metals in soil solids and breaks the SOM-mineral complex, resulting in a loss of SOM and an increase in DOM in solution.

Table 4-1 Equilibrium concentrations of metal cations in solution for phenanthrene sorption by soil

REC	Concentrations of REC/(g/L)	Concentrations of metal cations in solution/(mg/L)					
		Al	Cu	Fe	Mg	Mn	Zn
Alanine	0	1.01a	0.01a	0.61a	16.70b	2.37a	0.04a
	2	1.03a	0.07b	0.56a	16.42ab	2.39a	0.04a
	10	1.32a	0.22c	0.73a	14.99a	2.33a	0.03a
	40	11.83b	0.36d	6.54b	14.26a	2.34a	0.09b
Serine	0	1.01a	0.01a	0.61a	16.70a	2.37a	0.04b
	2	1.64b	u.d.	0.92b	16.81a	2.60b	0.02a
	10	1.68b	0.30b	0.95b	15.98a	2.55ab	0.04b
	40	9.70c	0.38b	5.25c	15.15a	2.55ab	0.10c
Fructose	0	1.01c	0.01	0.61b	16.70a	2.37a	0.04b
	2	0.46a	u.d.	0.27a	16.37a	2.63b	0.01a
	10	0.69b	u.d.	0.41ab	16.15a	2.93bc	0.01a
	40	1.97d	u.d.	0.82c	15.12a	3.37c	0.03ab
Malic acid	0	1.01a	0.01a	0.61a	16.70a	2.37a	0.04a
	2	16.36b	0.05b	22.68b	20.79b	6.76b	0.14b
	10	34.43c	0.16c	54.39c	29.17c	10.18c	0.54c
	40	69.90d	0.32d	115.90d	39.85d	12.83d	0.66d

Note: REC-root exudate component. u.d.- under detection limit. Values in the same columns followed by the same letter for the same REC treatment with different REC concentration are not significantly different ($p < 0.05$).

Next, we directly detected and calculated the concentrations of DOM (C_{DOM}) in solution for phenanthrene sorption by soils, as shown in Figure 4-5. The C_{DOM} values significantly increased with increasing concentrations of test organic acid. It is possible that the increased DOM in solution in the presence of malic acid is derived from the broken "bridges" of metallic cations in SOM-mineral complexes due to metal dissolution. Consequently, SOM concentrations are reduced since DOM is eluted from the soil. As discussed previously, SOM was the predominant pool of hydrophobic organic chemicals in soil (Chiou et al., 1979). Partitioning into SOM is the dominant mechanism of NOC sorption by soils with SOM contents >0.1%, while soil clay minerals contribute negligibly to NOC sorption (Chiou et al., 1998). SOM serves as the primary sorbent for phenanthrene. As such, the decreased SOM content may inhibit phenanthrene sorption in soils.

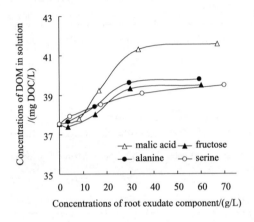

Figure 4-5 Equilibrium concentrations of DOM(mg DOC/L) in solution for phenanthrene sorption by soil

The environmental fate of phenanthrene in soil is dependent on its distribution among three phases: the aqueous, DOM, and soil solid phases (Seol and Lee, 2000; Gao et al., 2010b). The association of DOM with PAHs has been proposed as an important process that increases the solubility of PAHs in water (Celis et al., 1998; Ling et al., 2006). The interaction of PAHs with DOM may be more pronounced in the hydrophobic fractions, and the affinity of DOM to PAHs was controlled by the molecular properties of the contaminating PAHs (Totsche et al., 1997). Previously, the reported partition coefficient (K_{doc}) of phenanthrene with DOM was 8317.6 L/kg (Mott, 2002), which was generally much larger than their K_d between soil solids and water. Thus, phenanthrene is readily adsorbed by DOM in solution rather than by soil solids. Excluding the decreased SOM sorbent, it is possible that the inhibited sorption of phenanthrene by soil in the presence of malic acid may also be ascribed to the enlarged

DOM in solution (Figure 4-5).

The mechanism of the inhibited sorption of phenanthrene in soil by the addition of amino acids (alanine and serine) may be similar to that of organic acid, which was supported by the C_{Metal} and C_{DOM} data in Table 4-1 and Figure 4-5. This work revealed that fructose inhibited sorption of phenanthrene in test soils. In principle, fructose has limited ability to bind metal cations and break the "bridges" between soil minerals and organic matter in mineral-organic matter complexes. This was demonstrated by the weak influence of fructose on the C_{Metal} values in solution (Table 4-1). C_{DOM} values for phenanthrene sorption slightly increased in the presence of increasing concentrations of fructose in solution (from 0 g/L to 60.1 g/L), which resulted in inhibited phenanthrene sorption in soil due to the large K_{doc} value of phenanthrene with DOM in solution. The enhanced DOM may be derived from the release of SOM, but the mechanism involved is still unclear.

One notes that, excluding the changed SOM in soil solid and DOM in solution, root exudate and its components may also alter other chemical and biological properties of soils such as space structure of soil matrix, soil surface characteristics, and functional group composition and affinity ability of SOM. The observed apparent influences of RECs on the sorption of PAH in soil were conjunct function of above aspects. Unfortunately, little information is hitherto available on these REC-influenced soil properties. Thus, the mechanisms of the REC impacts on soil PAH availability still want further elucidation.

4.3 Impact of root exudates on PAH desorption from soils

Desorption from soil solid to soil water is the first step of NOCs to be absorbed by organisms, and the slow desorption of NOCs in soils has been a major impediment to the successful rhizoremediation of contaminated soils.

Here, we used phenanthrene and pyrene as representative PAHs, and sought to determine the effects of artificial root exudates (AREs) on the release of PAHs in three typical zonal soils in China. Rhizosphere model systems where soils are given AREs have been used to study the degradation of PAHs (Haby and Crowley, 1996). The advantage of this approach is that the quantity of added carbon can be controlled, and it is possible to avoid the interfering effects of inorganic nutrient depletion by roots (Hodge et al., 2000). The findings of this study may prove useful in the assessment of PAH-related risks to human health and the environment and rhizoremediation strategies at contaminated sites.

The composition of the AREs used containing 50 mmol/L glucose, fructose and sucrose, 25 mmol/L succinic acid and malic acid, 12.5 mmol/L serine, arginine, and cysteine (Joner et al., 2002). Brown-red soil, yellow-brown soil and red soil, all previously free of PAHs, were collected from the A horizon (0~20 cm). They are the typical zonal soils of China. Their basic properties are listed in Table 4-2. Soils were air-dried, ground and PAH-spiked (Gao et al., 2010). Treated soils were packed and NaN_3 solution (0.2%) was added to all microcosms to inhibit the microbial activities (Macleod and Semple, 2003). Then soils with 30% of the soil water holding capacity were aged for 0 day, 15 days, 30 days, 45 days, and 60 days at 25°C. The soils were sampled, sieved and ready for desorption experiments. A batch experiment was conducted to measure phenanthrene and pyrene release from aged soils. ARE solutions (20 mL, 0~1000 mmol organic carbon per liter) containing 0.01 mol/L KNO_3 were mixed with 1 g of treated soil in 30 mL glass centrifuge tubes. The tubes were shaken in the dark at 200 r/min on a gyratory shaker to reach equilibrium. The solution and soil were separated for analysis.

Table 4-2　Some properties of test soils

Soil	Location	f_{oc}/(g/kg)	Soil texture		
			Clay/%	Silt/%	Sand/%
Red soil	Yingtan, Jiangxi	4.94	36.8	22.5	40.7
Brown-red soil	Wuhan, Hubei	9.47	39.2	51.6	9.22
Yellow-brown soil	Nanjing, Jiangsu	14.3	24.7	61.9	13.4

Note: f_{oc} - Soil organic carbon content.

4.3.1　Desorption of PAHs from soils as a function of root exudate concentration

Desorption of phenanthrene and pyrene as a function of root exudate concentration (0~1000 mmol/L) is presented in Figures 4-6, 4-7 and 4-8. The addition of AREs significantly increased the release of phenanthrene and pyrene from test soils, and the higher the concentration of ARE, the larger desorption amounts of PAHs in soils was observed. For example, desorption of phenanthrene and pyrene in yellow-brown soil increased from 3.492 mg/kg and 1.562 mg/kg to 7.599 mg/kg and 2.526 mg/kg, respectively, when the ARE concentration increased from 0 mmol/L to 1000 mmol/L.

Similarly, after yellow-brown soil was aged for 60 days, phenanthrene and pyrene desorption also increased, from 2.066 mg/kg and 1.174 mg/kg without ARE to 2.793 mg/kg and 1.730 mg/kg with 1000 mmol/L ARE, respectively. The same tendency was observed in other soils and treatments. Thus, it can be safely concluded that the presence of AREs promotes PAH desorption from test soils, and that higher ARE concentrations elicit greater phenanthrene and pyrene desorption.

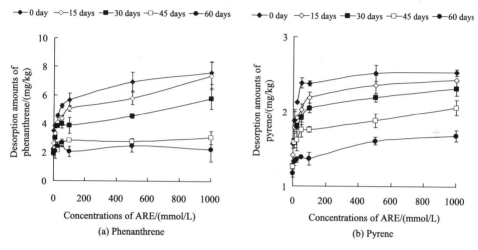

Figure 4-6 Desorption of phenanthrene and pyrene by ARE from yellow-brown soil aged for 0~60 days Error bars are standard deviations (SD)

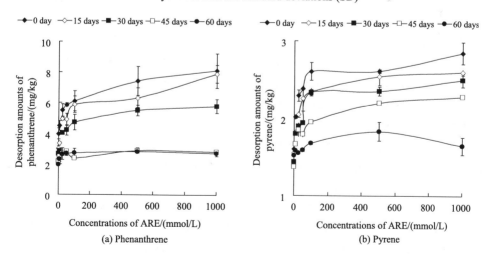

Figure 4-7 Desorption of phenanthrene and pyrene by ARE from brown-red soil aged for 0~60 days. Error bars are standard deviations (SD)

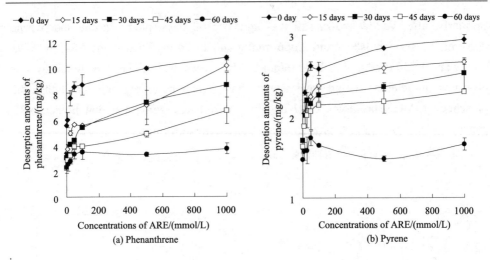

Figure 4-8 Desorption of phenanthrene and pyrene by ARE from red soil aged for 0~60 days. Error bars are standard deviations (SD)

4.3.2 PAH desorption by root exudates in different soils

SOM has an important impact on PAH desorption (Gao et al., 2006). In this study, although the addition of AREs significantly enhanced the desorption of phenanthrene and pyrene in all soils, desorption of PAHs was always lowest in yellow-brown soil (with the highest f_{oc} of 14.3 g/kg), regardless of the addition of AREs. Brown-red and red soils had relatively lower SOM contents (9.47 g/kg and 4.94 g/kg, respectively). For instance, the amounts of phenanthrene and pyrene desorbed from yellow-brown soil increased from 2.637 mg/kg and 1.416 mg/kg without AREs to 7.382 mg/kg and 2.429 mg/kg with 1000 mmol/L AREs, respectively. Desorption ratio denotes the amount of PAHs desorbed from contaminated soils to the initial total amount of PAHs in these soils; a higher ratio indicates more PAHs were desorbed from soils. The desorption ratio of phenanthrene and pyrene increased from 2.64% to 7.38% and from 4.72% to 8.10%, respectively. In brown-red soil, the phenanthrene and pyrene desorption ratios changed from 2.99% and 5.17% without AREs to 7.91% and 8.67% with 1000 mmol/L AREs, respectively. In red soil, which had the lowest SOM content, phenanthrene and pyrene desorption ratios changed from 3.36% and 5.38% to 10.1% and 8.88% with 1000 mmol/L AREs, respectively. The same trend was observed in the other treatments. From these results, we conclude that given the same ARE concentration, the desorption of PAHs was lowest in yellow-brown soil, moderate in brown-red soil, and highest in red soil. This is the opposite of the soils' f_{oc} content (yellow-brown-soil > brown-red soil > red soil). This trend might be observed because the progressive diffusion of phenanthrene and pyrene into the micropores of the organic fraction reduces the total desorbable amount; thus, phenanthrene and pyrene desorption was highest in red soil because it has the lowest organic matter content.

4.3.3 Effects of soil aging on PAH desorption by root exudates from soil

Phenanthrene and pyrene desorption in the presence of AREs with aging (incubation from 0 day to 60 day) is also shown in Figures 4-6, 4-7 and 4-8. Desorption decreased markedly over time for all treatments, with or without the addition of AREs. Freshly spiked soil (0 day) had the greatest desorbed fraction for the compounds studied. For example, in yellow-brown soil at day 0 mg/kg, 7.599 mg/kg and 2.526 mg/kg of phenanthrene and pyrene were desorbed, whereas only 2.214 mg/kg and 1.679 mg/kg were desorbed, respectively, if phenanthrene and pyrene were aged for 60 days in the presence of AREs at 1000 mmol/L. The desorption ratio in yellow-brown soil was 3.49%~7.60% for phenanthrene and 5.2%~8.43% for pyrene at day 0, compared to 1.93%~2.21% and 3.91%~5.60%, respectively, at day 60. If we take the amounts of phenanthrene and pyrene desorbed from soils incubated 0 day as 100%, the amounts of phenanthrene and pyrene desorbed from yellow-brown soil, brown-red soil, and red soil after 60 days were 29.1% and 66.5%, 33.5% and 57.8%, and 34.6% and 56.3%, respectively, at the same ARE concentration. Thus, aging had a strong effect on PAH desorption; the desorption effect of AREs diminished with aging time.

Desorption of NOCs diminishes with time was also supported in literature (Chung and Alexander, 1998). This aging process, which represents a sequestration of molecules, occurs with a number of organic pollutants. Aging of phenanthrene, 4-nitrophenol, and atrazine in the laboratory confirmed that these compounds become less available over time to bacteria, earthworms, or both (Louchart and Voltz, 2007). In our study, soils with different concentrations of AREs also showed significant aging effects. The reduced desorption of phenanthrene and pyrene over the 60-day period might indicate that aging involves diffusion into soil micropores, partitioning into SOM, strong surface adsorption, or a combination of these processes (Hatzinger and Alexander, 1995). These processes lead to the desorbed phases of phenanthrene changing from loosely bound phases to strongly bound phases with increasing residence time. In other words, transformation of PAHs took place from loosely bound phases to more persistent phases with increasing residence time. Despite this redistribution, the addition of AREs can still desorb a portion of strongly bound PAHs from soils.

4.3.4 Desorption of different PAHs by root exudates in soil

A comparison of phenanthrene and pyrene desorption with the addition of AREs is shown in Figure 4-9 for red soil aged for 45 days. The increment rates (r) of the desorption amount of phenanthrene and pyrene by ARE were calculated according to the equation: $r=(C_i-C_o)/C_i$, where C_i is the desorption amount of test PAH in soil with a certain concentration of ARE; and C_o is the desorption amount without ARE addition. The higher r value indicates the more significant increment of PAH desorption by ARE. As expected, the amount of phenanthrene that could be desorbed in the soil-water system was much greater than that of pyrene in each experimental treatment; in other words, phenanthrene desorption in soils was promoted by AREs to a greater degree than pyrene desorption, as indicated by the higher r values of the former (Figure 4-9). This result might be due to phenanthrene's higher solubility in water, lower K_{ow}, and lower molecular weight, so that at the same ARE concentration, the desorption of phenanthrene is always greater. Result indicates that higher molecular weight PAHs with more benzene rings are more recalcitrant in soils. This was also supported by their microbial biodegrdation in soils (Sabate et al., 2006).

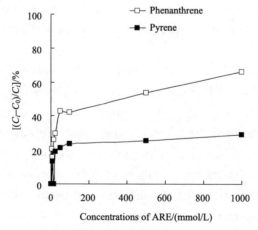

Figure 4-9   The increment rate (r) of the desorption amount of phenanthrene and pyrene by ARE in red soils aged for 60 days. r is calculated according to the equation: $r=(C_i-C_o)/C_i$, where C_i is the desorption amount of tested PAH in soil with a certain concentration of ARE, and C_o is the desorption amount in soil without ARE addition

4.3.5 Impact of root exudate components on PAH desorption in soil

Phenanthrene desorption in the presence of root exudate components (RECs) with ageing (incubation for 0~60 days) is shown in Figure 4-10 (Sun et al., 2012). Desorption decreased significantly over time for all treatments, with or without the addition of root exudates. Freshly spiked soil (0 day) had the largest desorbed fraction. For example, the desorption amounts of phenanthrene by control solutions without root exudates from soils aged for 0 day, 15 days, 30 days, 45 days and 60 days were 10.7 mg/kg, 9.76 mg/kg, 8.63 mg/kg, 7.67 mg/kg and 6.75 mg/kg, respectively. For soil treated with 59.4 g/L of alanine, if we take the amount of phenanthrene desorbed from soils incubated at 0 day as 100%, the amount of phenanthrene desorbed from soil after 15 days, 30 days, 45 days and 60 days of incubation was 85.7%, 79.1%, 65.4% and 62.2%, respectively. Thus, ageing had a strong effect on PAH desorption.

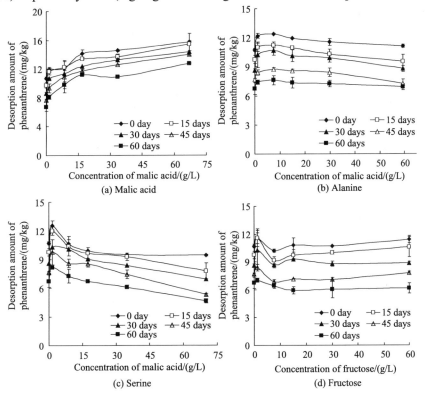

Figure 4-10 Desorption of phenanthrene from soils as a function of the concentration of the root exudate component. Legend with certain days denotes the PAH-spiked soil aged for these days. Error bars represent the standard deviation (SD)

Desorption of phenanthrene in soil, irrespective of ageing time, was significantly increased by organic acid (malic acid) at test concentrations (Figure 4-10). The higher the concentration of malic acids, the greater desorption of phenanthrene in soil. For instance, the desorption amounts of phenanthrene in soil aged for 45 days increased from 7.67 mg/kg to 9.34 mg/kg, 10.9 mg/kg, 11.7 mg/kg, 12.6 mg/kg and 14.0 mg/kg with increasing concentrations of malic acid from 0 g/L to 1.40 g/L, 8.38 g/L, 16.8 g/L, 33.5 g/L and 67.0 g/L, respectively. This was also observed in other soil samples and treatments.

In contrast, the influence of amino acids and fructose on PAH desorption differed from that of organic acid, as shown in Figure 4-10. At the test concentrations (0~60 g/L) of fructose, the desorption amounts remained almost constant, and the impact of fructose on phenanthrene desorption from soils was minimal. For amino acids, phenanthrene desorption increased initially and subsequently decreased with increasing concentrations of serine and alanine (0~70.1 g/L). This indicates that lower concentrations of amino acids increase PAH desorption, while higher concentrations of amino acids inhibit desorption. For example, the desorption amounts of phenanthrene in soils aged for 60 days by serine at concentrations of 1.46 g/L and 70.1 g/L were 8.15 mg/kg and 4.65 mg/kg, which were significantly 20.7% higher and 31.1% lower than that of the control (without serine), respectively.

4.3.6 Dissolved organic matter in soils with the addition of root exudates

Dissolved organic matter (DOM) is an important factor influencing desorption behavior of NOCs (Xing et al., 2008). DOM has been the subject of considerable interest in recent years because of its interaction with organic pollutants and thus its effects on the fate of these pollutants. A previous study examined the effects of DOM on the behavior of PAHs in a soil-water system (Cheng and Wong, 2006). To further elucidate the effects of AREs on PAH desorption in a soil-water system, we also determined the concentration of DOM in equilibrium solutions (C_{doc}), and how it changed with the addition of AREs. Because we wanted to evaluate the effects of AREs on the release of DOM from soils, the contribution of AREs to C_{doc} had to be eliminated.

As shown in Figure 4-11, C_{doc} increased with ARE concentration. In the three test soils, C_{doc} increased from 10.50~13.73 mg/L without AREs to 170.3~197.2 mg/L with AREs at 500 mmol/L, and tended to increase with increasing the ARE concentration. This result suggests that some of the inherent soil organic matter was eluted from the soil solids with the addition of AREs, resulting in their increased C_{doc}.

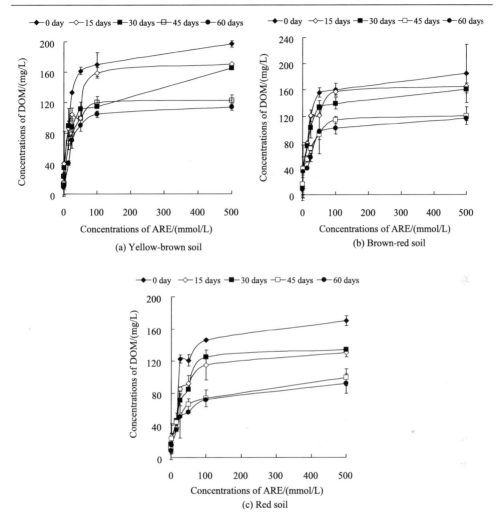

Figure 4-11   Concentrations of dissolved organic matters(mg DOC/L) in the test three soils with the addition of ARE. Error bars are standard deviations (SD)

It was notable that the desorption behavior of phenanthrene and pyrene might be influenced by increased DOM in the desorption solution with the addition of AREs (Figure 4-11), and the binding of phenanthrene and pyrene with DOM in solution would promote their desorption. This means that reduced SOM content in soils associated with the elution of inherent DOM leads to an increase in phenanthrene and pyrene desorption. The tendency of C_{doc} to vary as a function of ARE concentration in soils correlated positively with phenanthrene and pyrene desorption from soils, indicating that the ARE-induced DOM variation in aqueous solutions might account for the promoted desorption of phenanthrene and pyrene from soils in the presence of AREs. The association of DOM with PAHs has been proposed as an important process that results in solubility enhancement (Celis et al., 1998; Ling et al., 2006). Further, the interactions of PAHs with DOM are more pronounced in the hydrophobic fractions, and the affinity of DOM to PAHs is controlled by the molecular properties of PAHs (Raber and Kögel-Knabner, 1997). These findings indicate that PAHs tend to be adsorbed by DOM rather than other soil solids.

Moreover, the higher the soil f_{oc} content, the more DOM was eluted from the soil solids. For example, yellow-brown soil had the highest f_{oc} (14.3 g/kg), and its C_{doc} was 161.6 mg/L with the addition of 50 mmol/L ARE, which was 3.40% and 25.2% higher than that in brown-red and red soils, respectively, at the same ARE concentration. These results demonstrate that SOM plays an important role in PAH desorption; SOM serves as the primary sorbent for phenanthrene sorption. In our experiments, the observed increase in C_{doc} in ARE-supplemented soil compared to unspiked soil, in theory, would result in increased partitioning of phenanthrene to DOM in solution, and hence promote phenanthrene desorption in soil-water systems; thus, the higher DOM concentration in solution leads to a stronger promotion of PAH desorption by soil solids.

In addition, metallic cations complex with functional groups of organic molecules in solution, leading to the formation of "bridges" between soil solid surfaces and DOM in the aqueous phase (Saison et al., 2004). However, the addition of AREs, which can promote the dissolution of metallic cations and DOM from soils as mentioned above, broke the "bridges" between soil solid surfaces and DOM, and PAHs tended to be adsorbed by DOM, promoting the desorption of phenanthrene and pyrene from soils. Thus, the presence of AREs in soil-water systems promoted phenanthrene and pyrene desorption not only because of the increased DOM concentration (C_{doc}) in the equilibrium solution, but also owing to the decreased SOM content as a consequence of DOM in water through the broken "bridges" of metallic cations. However, it is still unknown which of these two mechanisms dominates the enhanced desorption of phenanthrene and pyrene by AREs.

4.4 Impact of root exudates on PAH availabilities in soils

The availability of NOCs in soil is a major factor limiting the efficiency of plant uptake and soil remediation. NOC risks measured according to their total concentrations in soil may be overestimated (Hatzinger and Alexander, 1995; Reid et al., 2000). When an NOC enters the soil, some portion will tightly combine with the soil organic matter and will not be released by aqueous solution. Organic solvents are therefore utilized to extract the available fractions of NOCs in soil (Kohl and Rice, 1998; Monteriro et al., 1999).

n-butanol has been reported as an appropriate extraction solvent to determine the availability of NOCs in soils (Liste and Alexander, 2002). Polycyclic aromatic hydrocarbons (PAHs) extracted by *n*-butanol significantly correlate with the PAHs absorbed by soil animals (such as earthworms) and plants, giving good predictions for their uptake by these organisms (Gomez-Eyles et al., 2010, 2011). Recently, Ling et al. (2009) used *n*-butanol extraction technique to determine the availability of phenanthrene and pyrene in rhizosphere soils. In general, *n*-butanol extraction has been widely recognized as a suitable technique for use in studies predicting the availabilities of PAHs in soil.

Here, we investigate the impacts of root exudates and their different components including organic acids, amino acids, and saccharides on the availability of pyrene as a representative PAH in soils utilizing *n*-butanol extraction technique. The simulated root exudates (SRE) used in this work consisted of 50 mmol/L fructose, 25 mmol/L malic acid, 25 mmol/L citric acid, 25 mmol/L oxalic acid, 12.5 mmol/L serine, and 12.5 mmol/L alanine, which were referred to Joner et al. (2002). A Yellow-brown soil was collected from Nanjing, China, and general properties were given in Table 4-2. Pyrene-spiked soils (TypicPaleudalfs) with present SRE or root exudate components (RECs) were incubated in microcosms for 30 days, and the available fraction of pyrene was determined using *n*-butanol extraction procedure (Liste and Alexander, 2002; Sun et al., 2013).

4.4.1 Impact of root exudates on *n*-butanol-extractable pyrene in soil

The concentrations of *n*-butanol-extractable pyrene in soil following the addition of SRE and a 30-day incubation are given in Figure 4-12. The extractable pyrene concentrations in sterilized and non-sterilized control soil without the addition of SRE were 5.14 mg/kg and 3.99 mg/kg, respectively. The addition of SRE significantly increased the amount of pyrene extracted from soils irrespective of sterilization, and the extractable pyrene also increased when more SRE were added to the soil. For example, the extractable pyrene concentrations in sterilized and non-sterilized soils with 20.62 g/kg SRE were 16.1 mg/kg and 12.60 mg/kg, respectively. These values were significantly much higher than those in the control soils, indicating the enhanced availability of pyrene in soils when SRE is present. The extractable concentrations of pyrene in sterilized soils were always significantly larger than those in non-sterilized soils irrespective of the addition of SRE, suggesting the microbial biodegradation of the extractable pyrene in non-sterilized soils.

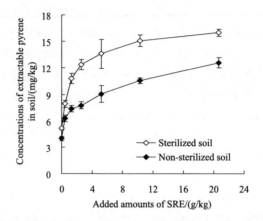

Figure 4-12   Concentrations of extractable pyrene as a function of the amount of simulated root exudates (SRE) added to the soil after a 30-day incubation. Error bars represent the standard deviations (SDs)

The enhancement ratio (r, %) of the extractable pyrene in soil by the addition of SRE or RECs was calculated as follows

$$r = (C_i - C_o)/C_o \times 100\%$$

where C_i and C_o are the extractable concentrations of pyrene in soil with and without the addition of SRE or RECs, respectively. The enhancement ratio (r, %) of the extractable pyrene in soil by the addition of SRE was obtained according to above equation. A bigger r value indicates a more significant enhancement of extractable pyrene in soil following the addition of SRE. The detected r values ($r_{detected}$) according to the equation are given in Figure 4-13. The $r_{detected}$ values increased straight ahead from 54% to 211% in sterilized soils and from 57% to 216% in non-sterilized soils with the increase of addition SRE from 0.43 g/kg to 20.62 g/kg, respectively. $r_{detected}$ values were always higher for sterilized soils than for non-sterilized soils.

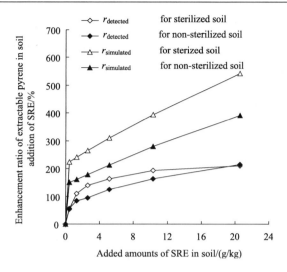

Figure 4-13   The enhancement ratio (r, %) of the extractable pyrene in soil following the addition of simulated root exudates (SRE); $r_{detected}$ and $r_{simulated}$ are the detected and simulated r values, respectively

4.4.2   Impact of root exudate components on the *n*-butanol-extractable pyrene in soil

The *n*-butanol extractable concentrations of pyrene in soils after a 30-day incubation with the addition of different RECs are displayed in Figure 4-14. Similar to the SRE, the addition of RECs clearly enhanced the extractable pyrene concentrations in soil irrespective of soil sterilization, and larger additions of RECs resulted in higher extractable concentrations of pyrene in tested soils. For example, the extractable concentrations of pyrene in sterilized soils with 0 g/kg, 0.79 g/kg, 2.36 g/kg, 4.73 g/kg, 9.46 g/kg and 18.91 g/kg oxalic acid were 5.14 mg/kg, 8.78 mg/kg, 11.09 mg/kg, 13.60 mg/kg, 14.33 mg/kg and 16.74 mg/kg, respectively, while those in non-sterilized soil were 5.14 mg/kg, 6.34 mg/kg, 8.82 mg/kg, 9.25 mg/kg, 10.16 mg/kg and 11.74 mg/kg, respectively. However, the amounts of extractable pyrene differed for different RECs. Generally, the extractable concentrations of pyrene in soils treated with RECs followed a descending order of organic acids (oxalic acid≥citric acid > malic acid) > amino acids (alanine > serine) > fructose. Organic acids were the most effective RECs for generating pyrene availability in soil, while fructose was the least effective. Moreover, the ability to make pyrene available clearly differed among amino acids or organic acids. The extractable pyrene concentrations in sterilized soils were always higher than those in non-sterilized soils, irrespective of the presence of RECs (Figure 4-14), again suggesting the microbial biodegradation of the extractable pyrene in non-sterilized soils.

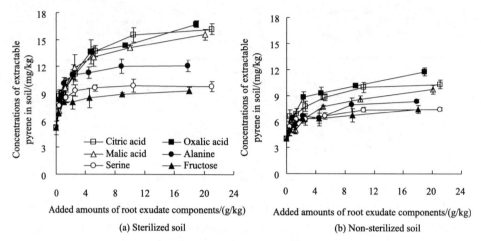

Figure 4-14 Concentration of extractable pyrene as a function of the amount of root exudates components (RECs) added in (a) sterilized and (b) non-sterilized soils after a 30-day incubation. Error bars represent the standard deviations (SDs)

The enhancement ratio (r, %) of the extractable pyrene in soil following the addition of RECs is shown in Figure 4-15. r values range between 26% and 194% and between 34% and 225% for non-sterilized and sterilized soils, respectively, indicating that more pyrene can be extracted from soils when RECs are present. Overall, r values followed a descending order of organic acids > amino acid > fructose. Organic acids always performed best in enhancing the availability of pyrene in soils, while fructose performed the worst. In addition, r values were clearly higher for sterilized soils compared to non-sterilized soils.

Here, the SRE consisted of organic acids, amino acids, and fructose. SRE concentrations can be calculated as follows

$$C_{exudate} = C_{citric} + C_{oxalic} + C_{malic} + C_{alanine} + C_{serine} + C_{fructose}$$

where $C_{exudate}$ is the SRE concentration in solution. C_{citric}, C_{oxalic}, C_{malic}, $C_{alanine}$, C_{serine}, and $C_{fructose}$ are the concentrations of citric acid, oxalic acid, malic acid, alanine, serine, and fructose, respectively, in solution. Suppose that the effects of SRE on pyrene availability in soil were a simple sum of the contribution from each REC at the test concentrations. Then the simulated r values ($r_{simulated}$) can be calculated as follows

$$r_{simulated} = r_{citric} + r_{oxalic} + r_{malic} + r_{alanine} + r_{serine} + r_{fructose}$$

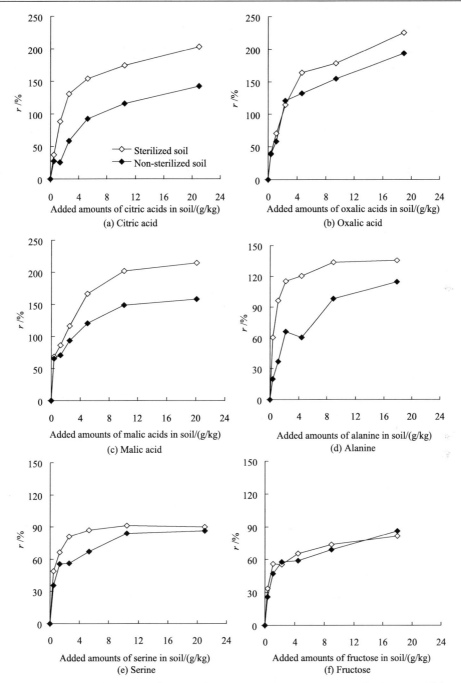

Figure 4-15 The enhancement ratio (r, %) of the extractable pyrene in soil following the addition of root exudates components (RECs). r was assessed as $r = (C_i - C_o)/C_o \times 100\%$, where C_i and C_o are the extractable concentrations of pyrene in soil with and without the addition of RECs, respectively

where $r_{simulated}$ is the simulated r value based on above equation. r_{citric}, r_{oxalic}, r_{malic}, $r_{alanine}$, r_{serine}, and $r_{fructose}$ (which can be obtained from Figure 4-15) are the enhancement ratios (%) of pyrene availability in soil by the addition of citric acid, oxalic acid, malic acid, alanine, serine, and fructose, respectively, at their respective concentrations of C_{citric}, C_{oxalic}, C_{malic}, $C_{alanine}$, C_{serine}, and $C_{fructose}$. The $r_{simulated}$ and detected r values ($r_{detected}$) are given in Figure 4-13. The $r_{simulated}$ values were always much higher than the actual values detected, irrespective of soil sterilization. For example, the $r_{detected}$ values for the addition of 5.15 g/kg and 20.62 g/kg root exudate in the sterilized soil were 164.4% and 211.2%, whereas the $r_{simulated}$ values were 309.8% and 542.6%, respectively. This indicates that the effects of SRE on pyrene availability in soil were not a simple sum of each REC contribution at the test concentrations, and that the RECs that are present in SRE may interfere with each other in terms of their influence on pyrene availability in soil.

It is obvious that microbial activities influence the available concentrations of NOCs in soils. Microbial catabolism has been reported in many studies, and is considered to be the principal mechanism for the removal of available NOCs, such as PAHs, from soils (Semple et al., 2003; Ling et al., 2009). The extractable fraction is the most available and biodegradable portion of NOCs for microbial catabolism in soils (Alkorta and Garbisu, 2001; Smalla et al., 2001). In non-sterilized soil, n-butanol-extractable pyrene is susceptible to microbial biodegradation. In sterilized soils, microbial activity was inhibited, and an increase in extractable pyrene residues was observed compared with non-sterilized soils. This demonstrates the dominant role of microbial biodegradation in decreasing the extractable pyrene fractions in natural soils (Sun et al., 2012).

4.4.3 Mechanisms by which root exudate and its components influence PAH availability in soil

Soil organic matter (SOM) is the main sink for PAHs in soil (Chiou et al., 1979, 1998). In soil, a linkage of metallic cations between soil minerals and SOM forms "bridges" of soil complexes (Saison et al., 2004). If the bridges are broken due to the release of SOM, metallic cations passing into solution, and mineral dissolution, the soil complexes will disappear. The functional groups of root exudates and their RECs, such as carboxyl and hydroxyl, may dissolve the metal cations and particularly the most common bridges of Al and Fe cations (White et al., 2003). Therefore, the bridges between the soil mineral surface and the SOM can be broken. Table 4-3 gives the concentrations of metal cations in solution, including aluminum (Al), calcium (Ca), iron (Fe), and magnesium (Mg), which are commonly found in soils following the addition of RECs. Clearly, the presence of organic acids dramatically increased the concentrations of metal cations in solution, and larger additions of RECs resulted in higher metal cation concentrations, indicating the significant release of metal cations from soil solids into solution. Because the connection of oxalic acid with Ca produces a precipitate, the concentrations of Ca for oxalic acid treatments were much smaller than those for control treatments without the addition of RECs.

Table 4-3 Equilibrium concentrations of metal cations in solutions extracted from pyrene-spiked soils aged for 30 days

Root exudate components	Concentrations of root exudate components/(g/L)	concentrations of metal cations in solution/(mg/L)			
		Fe	Mg	Ca	Al
CK	0	4.15±0.17	0.37±0.05	5.42±0.16	6.87±0.52
Citric acid	5	17.3±0.10	2.22±0.03	4.43±0.37	28.6±0.83
	10	23.3±0.58	2.84±0.06	6.42±0.05	37.1±0.48
	20	27.1±0.44	3.73±0.24	6.43±0.47	43.2±0.26
Malic acid	5	1.56±0.07	nd	4.28±0.10	2.85±0.22
	10	5.80±0.87	0.57±0.06	5.01±0.08	9.67±0.723
	20	13.9±0.26	2.11±0.08	5.94±0.07	20.8±0.010
Oxalic acid	5	4.88±0.05	0.64±0.37	4.68±0.06	8.31±0.78
	10	6.96±0.68	0.90±0.16	0.13±0.34	11.9±0.96
	20	13.6±0.99	2.07±0.77	nd	23.8±0.62
Serine	5	4.04±0.09	0.17±0.05	4.23±0.70	6.79±0.44
	10	4.42±0.81	0.31±0.03	4.54±0.34	7.47±0.19
	20	5.12±0.36	0.25±0.01	4.89±0.78	8.16±0.55
Alanine	5	3.77±0.36	0.74±0.12	5.42±0.08	6.18±0.56
	10	4.71±0.91	0.45±0.03	4.85±0.34	7.42±0.63
	20	5.13±0.11	0.37±0.03	4.89±0.25	8.24±0.64
Fructose	5	3.38±0.05	0.42±0.08	5.39±0.08	5.48±0.16
	10	4.39±0.20	0.51±0.03	4.58±0.07	7.72±0.34
	20	9.77±0.21	1.03±0.23	4.52±0.15	8.35±0.19

Note: CK means control treatment without REC addition. nd means not detectable. Data after "±" are standard deviations.

Due to the dissolution and release of metal cations from soil solids into solution, some SOM was eluted off from the soil solids following the addition of organic acids (Drever and Stillings, 1997), and became dissolved organic matter (DOM) in solution. This resulted in decreased SOM contents in soil solids and an increased DOM in solution. Because SOM serves as the main sink for pyrene, the reduced SOM content led to a higher pyrene availability in the soil (Weber et al., 2002; Li et al., 2005). The observed concentrations of DOM in solution extracted from soils following the addition of RECs are shown in Figure 4-16. The addition of organic acids dramatically increased the DOM concentrations in solution, suggesting a significant release of SOM from soil solids. Because the partition coefficient (K_{doc}) of PAHs in DOM is several magnitudes larger than their K_d values in SOM (Mott, 2002), the higher DOM concentrations in solution would result in lower fixation and higher availability of pyrene in soils (Lu et al., 2007; Sun et al., 2012, 2013). In addition, some studies have indicated that RECs, like organic acids, may dissolve the soil minerals and break the connection between soil minerals and SOM (Yang et al., 2001). This would also result in a release of SOM from soil solids into solution, and thus enhance the availability of PAHs in soils (Sun et al., 2012, 2013).

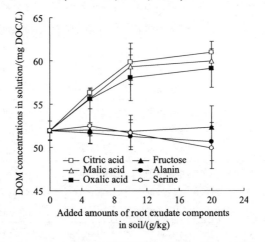

Figure 4-16 Equilibrium concentrations of DOM in solutions extracted from pyrene-spiked soils aged for 30 days. Error bars represent the standard deviations (SDs)

The test amino acids and fructose had little effect on the metal cation release (Table 4-3) and only minimally increased the DOM concentrations in solution (Figure 4-16). However, the addition of these RECs significantly enhanced the availability of pyrene in soils. The mechanisms involved in the enhanced availability of pyrene in soils following the addition of amino acids and fructose require further investigation.

Chapter 5 Low-molecular-weight organic acids (LMWOAs) influence the transport and fate of PAHs in soil

Low-molecular-weight organic acids (LMWOAs) occur widely in the soil environment as natural products of root exudates, microbial secretions, and decomposition of plant and animal residues. The most common LMWOAs identified to date in soils include oxalic, succinic, tartaric, fumaric, malic, and citric acids, among others (Kozdroj and Van Elsas, 2000; Nardi et al., 2005). LMWOA concentrations in soil are typically very low (10^{-3} mol/L to 10^{-5} mol/L), although high LMWOA concentrations occur in rhizosphere zones characterized by intense biological activity. However, under suitable environmental conditions, these acids may accumulate to levels toxic to plant growth (Kpomblekou-A and Tabatabai, 2003). Unlike long-chain fatty acids that may persist in soils for long periods of time, LMWOAs have a transitory existence; the amount of LMWOAs present in the soil at any time reflects a balance between synthesis and destruction processes controlled by microorganisms. By forming soluble complexes with polyvalent cations from rocks and minerals, organic acids also play an important role in the dissolution, transportation, and concentration of elements in the earth's surface as well as in soil formation and plant nutrition (Huang and Keller, 1972; Strahm and Harrison, 2008).

LMWOAs have been shown to disrupt the sequestering soil matrix, thereby enhancing the desorption and availability of organic pollutants like DDTs in soil (White et al., 2003). Consequently, it is expected that LMWOAs, in theory, may influence PAH transport and fate in soil environment. In this chapter, the impacts of LMWOAs on the sorption, desorption and availability of PAHs were investigated using batch techniques and microcosms. The elution of PAHs in soils by LMWOAs was evaluated using soil column experiments. In addition, the formation of a bound residue is generally considered to act as a soil detoxification process by permanently binding compounds into soil matrix, and the bioavailability of bound residues is the final endpoint for the risk assessment and regulatory management of organic chemicals in the soil environment (Richnow et al., 1995; Northcott and Jones, 2000). However, our investigation here revealed that LMWOAs dramatically enhanced the release of bound-PAH residues in soils.

5.1 LMWOAs-influence the PAH sorption by different soil particle size fractions

Sorption, as well as other chemical and biological processes involving organic pollutants, depends on the soil particle size (Denaix et al., 2001; Pernot et al., 2013). The particle size fractionation of soil has been well documented in recent decades (Schmidt et al., 1999a,b; Elfick et al., 2000; Christensen, 2001; Tang et al., 2009). The sorption of dissolved organic carbon (DOC) in soil shows that sorption mainly occurs on the silt and clay sized soil particles versus sand and particulate organic matter. Most (25% to 98%) of the DOC in soils is associated with the fine fraction (<53 μm) which consists of silt and clay-sized organo-mineral associations (Jagadamma et al., 2014). However, as to the PAHs, little published information is available on the sorption of PAHs by different particle size fractions of soil, and the function of root exudates as well as organic acids involved is under investigation. Understanding this issue is of paramount importance to the chemical processes, availability and microbial biodegradation of PAHs in rhizosphere soils.

Here, we seek to examine the impacts of LMWOAs on the sorption of phenanthrene (a representative PAH) by different particle size fractions of a test soil (Gao et al., 2015). The mechanism of the LMWOA-influenced sorption of phenanthrene by test soil solids is discussed based on the observed sorption of organic acid, the dissolution of metal cations and minerals in soil, and the competition from dissolved organic matter in solution that were released from soil solids. The findings of this work would be useful in the assessment of PAH availabilities and risks in the rhizosphere, and may be instructive in food safety and rhizoremediation strategies for contaminated sites.

A typical zonal soil (Typic Paleudalf) with no-detectable PAHs was collected from the A (0~20 cm) horizon in Nanjing, China. After air-drying and removal of stones and visible plant material, the soil samples were sieved through a 250 μm mesh. Then the soil samples with particle size smaller than 250 μm, named soil fine fraction (SFF), were obtained and ready for further particle-size fractionation. The basic properties of the soil samples are listed in Table 5-1. Citric and malic acids were selected as they commonly exist in soil as major components of the rhizospheric microenvironment and plant exudates (Gao et al., 2010b). A batch experiment was conducted to determine phenanthrene sorption in a soil solid-water system (Chiou et al., 1998; Ling et al., 2009; Gao et al., 2010b). Citric or malic acid solution containing 0.01 mol/L KCl and 0.05% Sodium azide (NaN_3) was mixed with 0.05 g samples of the SFF or its different particle size fractions in 20 mL glass centrifuge tubes with Teflon-coated closures. Concentrated methanol stock solutions were prepared with phenanthrene, and then 10 μL of phenanthrene stock solution in methanol were added to the above batch vials containing soil and aqueous solution. Kinetic experiment revealed that the phenanthrene sorption by test solids approached the equilibrium state in 24 h irrespective of the organic acid addition. The tubes were shaken under dark conditions at 25℃ and 200 r/min for 24 h to achieve equilibration. Then the solution and soil were separated for analysis.

Table 5-1 Some basic properties of the different soil particle size fractions

Soil samples (different particle size)	Content of different particle size fractions in SFF/(g/kg)	pH value (Soil : H_2O=1 : 2.5)	CEC /(cmol/kg)	Soil organic carbon content /(g/kg)	BET-N_2 surface area /(m^2/g)
soil fine fraction (<250 μm)	–	6.12	24	14.0	25.2
Fine sand (50~250 μm)	67.4	6.42	24	56.2	23.9
Silt (5~50 μm)	538.6	6.37	11	4.66	11.6
Coarse clay (1~5 μm)	323.4	6.44	60	23.9	59.0
Fine clay (0.1~1 μm)	70.4	6.78	66	21.9	80.2

Note: SFF, soil fine fraction. CEC, cation exchange capacity.

5.1.1 Fractionation protocol of different soil particle size fractions

Soil particles span a large size range, varying from stones and rocks down to submicron clays (<1 μm). The particle size distribution and chemical properties of soils have been the subject of numerous studies, and the environmental processes affecting organic pollutants are known to depend on the soil particle sizes (Christensen, 2001; Denaix et al., 2001; Tang et al., 2009). Various systems of size classification have been used to define arbitrary limits and ranges of soil-particle size. For example, the Soil Science Society of America adopts the United States Department of Agriculture

(USDA) classification. Geologists and geomorphologists typically use the Wentworth classification scheme (Wentworth, 1922) and that of Folk and variations of the Folk scheme (Prothero and Schwab, 1996). In this investigation, the separation of soil particles was performed according to the protocol developed by Institute of Soil Science, Chinese Academy of Sciences (ISSCAS, 1980), and this protocol has widely been used in China (Tang et al., 2010; Gao et al., 2015). Laser granulometry confirmed the suitability of the fractionation method used for the particle size distributions in this investigation.

The separation of soil particles was performed by the following protocol. All procedures were conducted in triplicate, and the protocol was completed within 2 days. Briefly, SFF (50 g) was treated ultrasonically at 25 ℃ for 10 min at a soil/water ratio of 1 : 20 to disperse macroaggregates. The ultrasonication protocol was referred to literature (Tang et al., 2009), and the ultrasonic energy used for dispersion was 500 J/mL (Schmidt et al., 1999a,b). The fine sand fraction (50~250 μm) was removed by wet sieving through a 270 mesh (53 μm), and the suspension was then transferred to a 2-L glass beaker. Deionized water was added to completely disperse the sample; it was then mixed thoroughly and left to sediment for 1 h and 14 min (at 20 ℃). The sedimentation time was calculated according to Stokes' law. The upper layer of liquid (0~10 cm from the surface) was siphoned off, and sedimentation and siphoning was repeated 15~20 times until the upper layer liquid appeared clear. All of the siphoned liquid was combined, and the sediment was separated to obtain the silt fraction (5~50 μm). The siphoned liquid after the separation of the silt fraction was centrifuged at 1500 r/min for 5 min to separate the coarse clay fraction (1~5 μm). The supernatant was collected and centrifuged at 4500 r/min for 54 min to separate the fine clay fraction (0.1~1 μm). From this procedure, four size fractions (50~250μm, 5~50μm, 1~5μm, and 0.1~1 μm) of soil particles were obtained. All four fractions were freeze-dried and stored in polyethylene bottles prior to further analysis. The fine sand, silt, coarse clay, and fine clay content in the SFF were 67.38 g/kg, 538.60 g/kg, 323.44 g/kg, and 70.40 g/kg, respectively, and their basic properties including pH, CEC, organic carbon content, and surface area are presented in Table 5-1. One notes that the sum of the four obtained fractions does not yield 100% of SFF. This is because of nanocolloidal particles (<0.1 μm) left after the fractionation protocol, but the amounts of nanocolloidal particles are very small. To our knowledge, little information in literatures could be found about the organic pollutant sorption by soil nanocolloidal particles due to the experimental difficulties.

The particle size distribution of the four soil fractions as determined by laser granulometry is plotted in Figure 5-1. Table 5-2 summarizes the statistical parameters that were used to describe the volume distributions of each fraction, in which D10 represents the size interval that includes the first 10% of the particle volume distribution, D50 the size interval that includes the first 50% of the volume distribution (i.e. the median particle size), and D90 the size interval that includes the first 90% of the volume distribution, respectively (Elfick et al., 2000). The fine sand fraction (50~250 μm) had a unimodal and well-defined size distribution, mostly in the range of 26.1~211 μm and with a mean diameter of 81.7 μm, indicating that wet sieving was effective in separating this fraction. The silt fraction (5~50 μm) had a mean diameter of 15.9 μm and showed a bimodal distribution with the primary mode in the range of 4.88~32.7 μm and a weak secondary distribution with a mode diameter of 0.08~4.19 μm. This effect may be attributable to the presence of a small amount of water dispersible soil aggregates in the silt fraction, which otherwise had a well-defined size distribution (Tang et al., 2009). Both the coarse clay (1~5 μm) and fine clay fractions (0.1~1 μm) had unimodal, symmetrical, and well-defined size distributions between 1.06~5.74 μm and 0.29~1.48 μm with mean diameters of 2.64 μm and 0.68 μm, respectively. These results indicate a good level of agreement between the particle size ranges determined by the particle analyzer and the particle size ranges of the fractionation protocol as shown in Table 2, confirming the suitability of the fractionation methods used for the particle size distributions in this investigation.

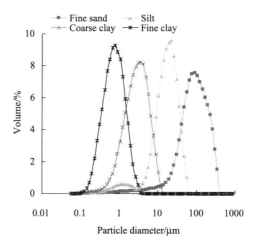

Figure 5-1   Particle size distribution of the test soil fractions

Table 5-2 Characterization of particle size distribution of the test soil fractions

Soil fractions	Particle size of the fractionation protocol/μm	D10	D50	D90
Fine sand	50~250	26.1	81.7	211
Silt	5~50	4.88	15.9	32.7
Coarse clay	1~5	1.06	2.64	5.74
Fine clay	0.1~1	0.29	0.68	1.48

Note: D10 represents the size interval that includes the first 10% of the particle volume distribution; D50 the size interval that includes the first 50% of the volume distribution (i.e. the median particle size); D90 the size interval that includes the first 90% of the volume distribution (Elfick et al., 2000).

5.1.2 PAH sorption by different soil particle size fractions

The sorption isotherms of phenanthrene by SFF and its four different particle size fractions could be significantly described using a linear distribution-type model (Figure 5-2(a)). The distribution coefficient (K_d, L/kg) of phenanthrene between soil and water according to the linear sorption model is expressed as

$$K_d = Q / C_w$$

where Q (mg/kg) is the amount of phenanthrene sorbed by the soil samples and C_w (mg/L) is the equilibrium concentration of phenanthrene in the aqueous phase. In this study, the value of K_d for phenanthrene sorption by SFF and its fine sand, silt, coarse clay, and fine clay fractions were given in Table 5-3. The K_d of a nonpolar contaminant (solute) is strongly dependent on the SOM content, and SOM serves as the primary sorbent for nonpolar contaminant sorption unless SOM content is extremely low (<0.1%) (Celis et al., 1998; Chiou et al., 1998; Gao et al., 2006). In this investigation, the K_d values were significantly positively correlated to the organic carbon contents of test soil solids (Figure 5-2(b)), indicating that SOM dominates phenanthrene sorption by soil solids. As such, silt fraction with the smallest organic carbon content (4.66 g/kg) consistently showed a lowest sorption potential as compared to other fractions and test SFF.

The values of the corresponding carbon-normalized distribution constant (K_{oc}, L/kg) were also obtained. K_{oc} can be calculated as

$$K_{oc} = K_d / f_{oc}$$

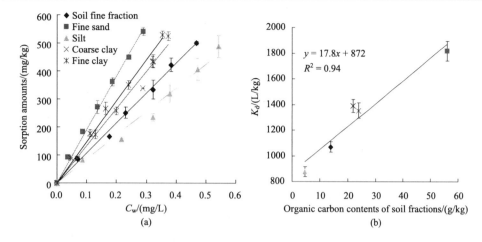

Figure 5-2 Sorption of phenanthrene by different soil particle size fractions. (a) Sorption isotherms. (b) Correlation of K_d with organic carbon contents of soil fractions. Error bars are standard deviations

Table 5-3 The distribution coefficient (K_d, L/kg) and the carbon-normalized distribution constant (K_{oc}, L/kg) of phenanthrene between soil solids and water

Soil samples (different particle size)	K_d/(L/kg)	lgK_{oc}
Soil fine fraction (< 250 μm)	1070	4.9
Fine sand (50~250 μm)	1817	4.5
Silt (5~50 μm)	877	5.3
Coarse clay (1~5 μm)	1352	4.8
Fine clay (0.1~1 μm)	1390	4.8

where f_{oc} is the fractional organic carbon content of SFF or its different particle size fractions (%, OC/100). The values of lgK_{oc} obtained for phenanthrene sorption by SFF and its fine sand, silt, coarse clay, and fine clay fractions varied from 4.5 to 5.3 with an average of 4.8, suggesting that except for the total contents of SOM, the characteristics of SOM may also play a role in phenanthrene sorption (Gao et al., 2006). It has been generally accepted that the nature or location of natural organic matter in soil affects the effective level of active organic matter to sorb non-ionic organic chemicals (NOCs) (Chiou et al., 1983; Murphy and Zachara, 1995). Here, such K_{oc} variations can be attributed to the nature and compositional differences in SOM.

5.1.3 Effects of LMWOAs on PAH sorption by different soil particle size fractions

The sorption of phenanthrene by SFF and its different particle size fractions was determined in the presence of citric and malic acids as representative LMWOAs. In all cases, sorption could be described well by a linear sorption isotherm. The K_d values for phenanthrene sorption by SFF and its fractions in the presence of citric and malic acids at concentrations ≤1000 mmol/L were obtained and are displayed in Figure 5-3. The K_d values followed a descending order of fine sand > fine clay > coarse clay > silt for the treatments with the addition of organic acids. As with the control treatment (without addition of organic acid), this order was also significantly correlated with the organic matter contents of the four soil fractions, suggesting the SOM-dominated sorption of phenanthrene by soil solids in the presence of organic acids.

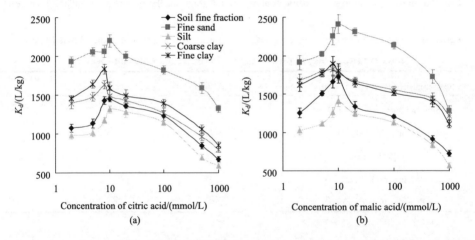

Figure 5-3 Sorption of phenanthrene by different soil particle size fractions as a function of the addition concentrations of (a) citric and (b) malic acid in solution

Similar trends were observed for phenanthrene sorption by both SFF and its four size fractions as a function of the concentrations of organic acids in aqueous solution. As shown in Figure 5-3, the K_d values for phenanthrene sorption by SFF and its four size fractions initially increased but then decreased as the concentrations of citric and malic acids increased (0~1000 mmol/L). C_{max}, that is, the concentration of organic acid corresponding to the largest K_d value, was observed at 10 mmol/L for SFF, fine sand, and silt fractions, and at 8 mmol/L for the coarse and fine clay fractions. The presence of citric and malic acids at lower concentrations (<100 mmol/L) generally promoted the sorption of the test PAH. In contrast, higher organic acid concentrations (>100 mmol/L) in the soil-water system clearly impeded the distribution of phenanthrene into solids irrespective of different particle size fractions (Figure 5-3). For example, the maximum K_d values of 2198 L/kg, 1318 L/kg, 1635 L/kg, and 1834 L/kg for phenanthrene sorption by fine sand, silt, coarse clay and fine clay at the C_{max} of citric acid were 21%, 51%, 21% and 32% higher, respectively, than their corresponding control K_d values, and the maximum K_d values of 2411 L/kg, 1411 L/kg, 1820 L/kg, and 1892 L/kg at the C_{max} of malic acid were 33%, 61%, 35%, and 36% higher, respectively, than the control values. In contrast, the minimum K_d values for phenanthrene sorption by fine sand, silt, coarse clay and fine clay at 1000 mmol/L citric acid were 27%, 33%, 41% and 39%, and those at 1000 mmol/L malic acid were 29%, 35%, 10% and 20% lower than the control values, respectively.

In recent years, the sorption of PAHs by soils in the presence of exotic organic acids was documented. Ouvrard et al. (2006) reported that phenanthrene sorption by soil (0.1 g soil with 20 mL solution) was enhanced as malic acid presented at low concentrations (0~20 mmol/L), which was in agreement with our observations. In contrast, Ling et al. (2009) revealed that phenanthrene sorption by soil (0.5 g soil with 25 mL solution) was clearly impeded by the addition of citric acid (0~1000 mmol/L). In this investigation, by using a smaller ratio of soil solids to solution (0.05 g soil with 10 mL solution), we observed that citric and malic acids at lower concentrations (<100 mmol/L) promoted the sorption of the PAH tested, while higher organic acid concentrations (>100 mmol/L) inhibited its sorption (Figure 5-3).

5.1.4 Mechanisms of LMWOA-influenced PAH sorption by different soil particle size fractions

The mechanisms involved in the LMWOA-influenced PAH sorption by soil may be due to the cumulative sorption of organic acid, the dissolution of metal cations and minerals in soil, and competition from DOM in solution, which are also discussed similarly in "4.2.2 and 4.4.3".

In this study, the sorption of organic acids, citric acid as an example, in the soil-water system was observed, and the levels of sorption as a function of the equilibrium concentrations of citric acid are shown in Figure 5-4. The citric acid added during the experiments was captured by SFF and its different particle size fractions, and 6%~32% citric acid over the range of test concentrations (0~1000 mmol/L) was sorbed. Freundlich isotherm equations were applied to interpret the experimental data for citric acid sorption. The Freundlich isotherm can be described by the following equation

$$Q_e = K_F \cdot C_e^n$$

where Q_e (mmol/g) is the amount of citric acid sorbed by the soil solid at equilibrium; C_e (mmol/L) is the equilibrium concentration of citric acid in solution; K_F (mL/g) and n (dimensionless) are Freundlich adsorption isotherm constants, indicative of the extent of adsorption and the degree of nonlinearity between the solution concentration and adsorption, respectively. The parameters of these sorption isotherms are listed in Table 5-4. It is clear that the citric acid sorption by test solids could be well simulated by the Freundlich isotherm model. The capacity of citric acid sorption, as displayed by the K_F values in Table 5-4, followed a descending order of fine clay > coarse clay ≫ silt fine > fine sand, which was negatively correlated to the particle size of these soil fractions.

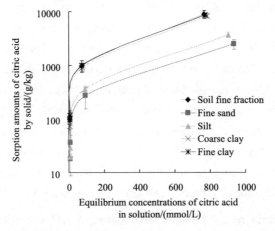

Figure 5-4 Sorption of citric acid by different soil particle size fractions. Error bars are standard deviations

Table 5-4 Parameters of Freundlich isotherms for the sorption of citric acid by test soil solids

Soil samples (different particle size)	K_F/(mL/g)	n	R^2
Soil fine fraction (< 250 μm)	100.1	0.9214	0.9991
Fine sand (50~250 μm)	11.42	0.8120	0.9842
Silt (5~50 μm)	12.82	1.0828	0.9986
Coarse clay (1~5 μm)	72.66	0.9600	0.9989
Fine clay (0.1~1 μm)	103.7	0.9141	0.9998

Note: K_F (mL/g) and n (dimensionless) are Freundlich adsorption isotherm constants, indicative of the extent of adsorption and the degree of nonlinearity between the solution concentration and adsorption, respectively.

The enhanced sorption of phenanthrene by SFF and its different particle size fractions in the presence of organic acid at the lower concentrations (<100 mmol/L) could be explained by the adsorption of organic acid on the soils itself thereby constituting a new range of sorption sites for phenanthrene. This mechanism was suggested by the findings of Gao et al. (2006) and Ouvrard et al. (2006). Organic acid sorption to soil solids may increase the bulk SOM content and hence provide new sorption sites. As a result, such "cumulative sorption" would increase the soil's capacity for taking up phenanthrene and promote its sorption by the soil and its different particle size fractions (Ouvrard et al., 2006; Gao et al., 2006, 2007, 2010b).

Metallic cations in soil form complexes with functional groups of soil organic molecules, and this association leads to the formation of "bridges" between minerals and the SOM in soil (Saison et al., 2004; Gao et al., 2012). LMWOAs are found in the rhizosphere as monocarboxylic, dicarboxylic, and tricarboxylic acids containing saturated or unsaturated carbon and hydroxyl groups. The binding of LMWOAs with metal cations has been well documented in previous studies (Saison et al., 2004; Kong et al., 2013). In our study, the added organic acids dissolved the metal cations in SFF and its fractions (Table 5-5), hence breaking the bridges between the soil mineral and organic matter. The release of metal cations from soil solids to solution breaks the "bridge" between minerals and the SOM in soil, and as a consequence, the organic matter complexed with soil minerals through the released metal cations becomes DOM in solution (Gao et al., 2010b, 2012), resulting in decreased SOM solid content and increased DOM in solution. On the other hand, the dissolution of some soil minerals by root-secreted organic acids has also been reported (Drever and Stillings, 1997), which also theoretically breaks the mineral-organic matter complex, reduces the SOM in soil, and increases the DOM in solution.

Since SOM serves as the primary sink for phenanthrene sorption (Chiou et al., 1998), the decreased SOM content due to the addition of organic acid may result in inhibition of phenanthrene sorption by the test SFF and its fractions. However, the calculated decrease in SOM content in solid samples following the addition of citric acid compared to the controls were less than 3%, which was negligible when compared to the variation of K_d (Figure 5-3). This suggests that decreases in the SOM content have a minor effect on the inhibited phenanthrene sorption following the addition of organic acid at higher concentrations (>100 mmol/L).

Table 5-5 Equilibrium concentrations of some metal cations in aqueous solution for phenanthrene sorption by different soil particle size fractions

Soil samples (different particle size)	C_{citric}/(mmol/L)	Concentrations of metal cations in solution/(mg/L)			
		Al	Ca	Fe	Mg
Soil fine fraction (< 250 μm)	0	0.15a	11.0a	0.52a	2.80a
	8	3.72b	13.5b	9.11b	3.28b
	10	4.46bc	14.1b	10.3b	3.30b
	100	10.7d	14.3b	22.6c	5.22c
Fine sand (50~250 μm)	0	0.23a	11.8a	0.56a	2.85a
	8	4.30b	14.1b	9.65b	3.35b
	10	4.77b	15.5c	10.1b	3.42b
	100	10.7c	15.5c	23.8c	5.15c
Silt (5~50 μm)	0	0.57a	4.28a	0.61a	1.19a
	8	1.56b	5.12ab	2.74b	1.19a
	10	1.63b	5.30b	2.97b	1.23a
	100	2.41c	6.19c	4.68c	1.40b
Coarse clay (1~5 μm)	0	0.37a	14.6a	0.52a	4.12a
	8	5.72b	15.8ab	9.24b	4.35a
	10	6.12b	16.4b	11.3c	4.48ab
	100	12.8c	17.3c	17.8d	6.30b
Fine clay (0.1~1 μm)	0	1.21a	16.7a	0.93a	5.53a
	8	8.43b	22.9b	8.76b	6.59b
	10	9.35b	23.9bc	10.2b	6.99b
	100	18.0c	24.8c	20.2c	8.51c

Note: C_{citric}, the concentration of citric acid in solution. Different lowercase letters indicate significant differences ($p < 0.05$) among citric acid treatments for the same metal cation and soil solid.

The presence of DOM in solution inhibits PAH sorption by soils (Gao et al., 2007). The association of DOM with PAHs in solution has been proposed as an important process that results in enhanced solubility (Celis et al., 1998; Ling et al., 2005, 2006; Barriuso et al., 2011). The affinity of DOM to PAHs is controlled by the molecular properties of the PAHs (Raber and Kögel-Knabner, 1997). The reported partition coefficient (K_{doc}) of phenanthrene with DOM in aqueous solution is 8317.6 L/kg (Mott, 2002). Clearly, the reported K_{doc} value was much higher than the observed

K_d values for phenanthrene sorption by soil solids (Figure 5-3); therefore, phenanthrene tends to be captured by DOM in solution rather than by soil solids. In this study, the concentrations (C_{DOC}) of the dissolved organic carbon (DOC) released from soil solids in solution for phenanthrene sorption are shown in Figure 5-5. This portion of DOC was determined by eliminating the contribution of organic acid and phenanthrene to the total DOC concentration in solution. The observed DOC may have been derived from the breakdown of mineral-SOM complexes because of the dissolution of metal cations and minerals by organic acid. The inhibited sorption of phenanthrene by the addition of organic acid at higher concentrations (>100 mmol/L) may primarily be ascribed to the reason that organic acid eluted off DOM from the soil solids, and the DOM at higher concentrations in solution inhibited the sorption of phenanthrene by soil solids.

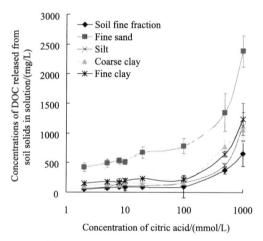

Figure 5-5 Concentrations of dissolved organic carbon (DOC) released from soil solids in solution for phenanthrene sorption by different soil particle size fractions. Error bars are standard deviations

Note that the apparent influence of LMWOAs on the sorption of phenanthrene by SFF and its different particle size fractions may actually be due to a combination of factors. Excluding the cumulative sorption of organic acid, the dissolution of metal cations and minerals in soil, and the competition from DOM in solution, LMWOAs may also alter other chemical and biological properties of soil solids such as soil surface characteristics, the space structure of the soil matrix, and the composition and other properties of SOM (Sun et al., 2012). In this investigation, the addition of organic acids in soil-water systems would result in the reduced pH values and enlarged ionic strength in solution. Since SOM serves as the dominant sorbent of nonpolar contaminants for soils with relative high organic carbon contents (Chiou et al., 1979, 1998), the changed pH values and ionic strength in solution due to organic acid addition would theoretically influence the properties of SOM and eventually affect the sorption of phenanthrene by SFF and its fractions. Unfortunately, little information is hitherto available regarding these LMWOA-influenced SOM properties. The mechanisms controlling the impacts of LMWOA on PAH sorption by soil and its fractions therefore require further investigation.

5.2 LMWOAs enhance the PAH desorption from soil

Desorption of hydrophobic organic contaminants including PAHs is believed to have dominant effect on their bioavailability and fate in soils (Pignatello and Xing, 1996). Because bioavailability is a factor often limiting the degradation of PAHs in soils over time, enhanced desorption in rhizosphere could be adopted as an approach to promote the success of rhizoremediation. Amendment of soil with LMWOAs is also easily implemented in the field (Subramaniam et al., 2004). Recently, Gao et al. (2010) used artificial root exudates (AREs) as amendments, and observed a marked increase of desorption of phenanthrene and pyrene from soils. LeFevre et al. (2013) showed that both artificial and plant-harvested root exudates enhanced naphthalene desorption from soils, providing an abiotic contribution to the "rhizosphere effect" for degradation of naphthalene. Among the many substances in the exudates, LMWOAs are major components, and occur widely in soils environment (Jones, 1998; Ling et al., 2009). They have been shown to ably disrupt the sequestration of organic compounds in soil matrices, thereby enhancing desorption and bioavailability of dichlorodiphenyldichloroethylene (DDE) and dichlorodiphenyltrichloroethane (DDT) (White et al., 2003). It is therefore very likely that LMWOAs can facilitate the desorption of sequestrated PAHs in soils.

Here, we seek to examine the effects of LMWOAs on desorption of PAHs from PAH-spiked soils and soils collected from a PAH-contaminated site. Citric, oxalic and malic acids were selected as representative LMWOAs because they commonly exist in the rhizosphere, and are major components of plant exudates (An et al., 2011; Gao et al., 2011). The results from the present study could help to develop sound remediation strategies for PAH-contaminated sites by enhancing the PAH bioavailability in soils.

5.2.1 LMWOA-enhanced desorption of PAH from PAH-spiked soil

Brown-red soil, yellow-brown soil and red soil, all previously free of PAHs, were collected from the A horizon (0~20 cm). Their basic properties are listed in Table 4-2. Phenanthrene-spiked soils were prepared according to the method reported by Gao et al. (2010b). The spiked soils with 30% of the soil-water holding capacity and were aged for 0~60 days in microcosms at 25℃. Then soils were freeze-dried, sieved and ready for desorption experiments. A batch experiment was conducted to measure phenanthrene desorption from aged soils. An aqueous solution (25.0 mL) containing 0 mmol/L to 1000 mmol/L organic acid, 0.01 mol/L KCl and 0.5 g/L NaN_3 were mixed with 0.50 g of the soil in 30-mL glass centrifuge tubes. The tubes were shaken at 200 r/min on a rotator in the dark at 25℃ to reach equilibrium. The solution and soil were separated for analysis.

Phenanthrene desorption curves obtained from the various LMWOA extractions, including the untreated (day 0) soil sample and samples aged for 15 days, 30 days, 45 days, and 60 days were illustrated in Figures 5-6, 5-7 and 5-8. Also shown in this figure are the desorption amounts of phenanthrene from three tested soil samples in the presence of LMWOAs with concentrations ranging from 0 mmol/L to 1000 mmol/L. The desorption of phenanthrene increased incrementally with increasing LMWOA concentrations from 0 mmol/L to 1000 mmol/L. For example, without organic acid present in the yellow-brown soil, the phenanthrene desorption increased from 5.44 mg/kg at baseline to 10.88 mg/kg with the addition of citric acid to 1000 mmol/L. The same tendency was observed with oxalic and malic acids, with increases of 5.95 mg/kg and 5.80 mg/kg, respectively.

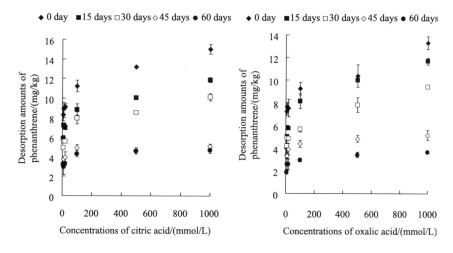

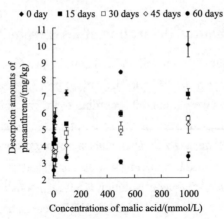

Figure 5-6 Desorption of phenanthrene from red soil aged for 0~60 days as a function of organic acid concentrations. Error bars are standard deviations (SD)

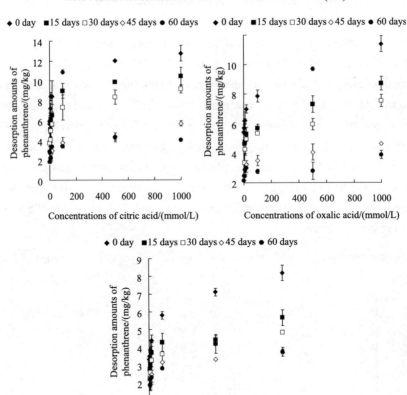

Figure 5-7 Desorption of phenanthrene from brown-red soil aged for 0~60 days as a function of organic acid concentrations. Error bars are standard deviations (SD)

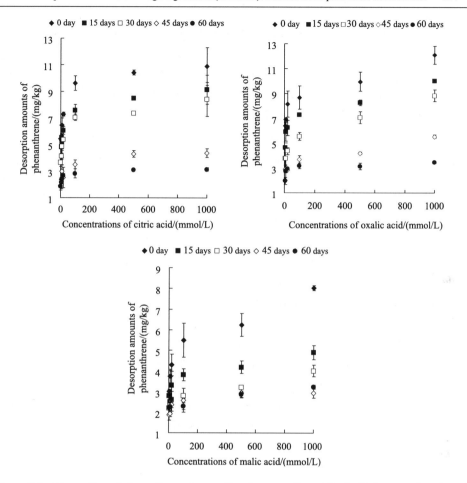

Figure 5-8 Desorption of phenanthrene from yellow-brown soil aged for 0~60 days as a function of organic acid concentrations. Error bars are standard deviations (SD)

The desorption amounts of phenanthrene varied markedly for various LMWOAs. The effect of organic acid concentration on the phenanthrene desorption amounts for all soil types was always larger for citric acid than for oxalic or malic acids. For example, the desorption amount of phenanthrene in soil 1 supplemented with 1000 mmol/L citric acid was 15.02 mg/kg, whereas those in soil supplemented with oxalic and malic acids were 13.30 mg/kg and 10.00 mg/kg (88.5% and 66.6% of citric acid), respectively. This variation may result from the chemical structure differences among the acids; citric acid is a ternary organic acid, so it has a stronger effect than either oxalic or malic acids, both of which are binary organic acids. Regardless, LMWOAs are efficient at promoting the desorption amounts of phenanthrene. These results imply that the desorption behavior of phenanthrene is related to organic acid type and concentration, although the soil properties may be different.

As mentioned, SOM may also influence the soil desorption of phenanthrene. Figure 5-6, Figure 5-7 and Figure 5-8 illustrate the phenanthrene desorption amounts with the addition of citric acid versus the control soil samples that were positively correlated with their f_{oc}. The desorption amounts of phenanthrene in yellow-brown soil decreased from 10.88 mg/kg at the initial soil sampling time (0 day) to 3.16 mg/kg when allowed to age for 60 days, where as the amounts in the red soil and brown-red soil decreased from 15.02 mg/kg and 12.76 mg/kg to 4.67 mg/kg and 4.04 mg/kg, respectively. These values corresponded to their f_{oc}, in that a higher f_{oc} value represented a lower desorption amount. This observation may be due to the progressive diffusion of phenanthrene into the microporosity of the organic fraction, thus reducing its desorption amount. Therefore, the amount of phenanthrene desorbed in red soil is more significant than in brown-red and yellow-brown soils, as the former has a lower SOM content than the other soil types.

The desorption amounts of phenanthrene in the presence of LMWOAs with varied aging time (0~60 days) are shown in Figures 5-6, 5-7 and 5-8. The amounts of phenanthrene desorbed from the soil samples decreased as a function of incubation time, from 0 day to 60 days. In freshly spiked soils, 5.44~10.88 mg/kg phenanthrene was desorbed from the yellow-brown soil, and phenanthrene desorption was highest in the presence of 1 mol/L LMWOAs. In contrast, only 1.87~3.16 mg/kg phenanthrene was detected after the soil was allowed to age for 60 days. A similar tendency was observed in the other treatment groups. This observation indicates that the desorption amount of phenanthrene diminishes over the incubation of soil samples. These results may be explained by the slow molecular diffusion of contaminants into micropores within the sorbent phases (organic matter), as described in the literature (Pignatello and Xing, 1996; Glover et al., 2002).

5.2.2 LMWOA-enhanced desorption of PAHs from soils collected from a PAH-contaminated site

To test if LMWOAs are will effectively enhance desorption of PAHs from contaminated soils in the real environment, a soil sample was collected from a known PAH-contaminated site (Ling et al., 2015), and examined for the desorption using the approaches similar to the previous study (Ling et al., 2009). Twenty PAH-contaminated soil sub-samples were randomly collected from a site near a petrochemical manufacturing plant in Nanjing, China, and mixed together to form a composite sample. The soil was classified as Typic Paleudult (IUSS), and was air-dried, and sieved through a 0.3 mm mesh. The soil pH was measured as 5.87 (soil mass: water volume ratio = 1 : 2.5). Soil organic carbon (SOC) content was 13.6 g/kg and texture was 26.3% clay, 13.0% sand and 60.7% silt. The PAH concentration in the soil was 1552 ± 315 µg/kg (n = 5) for phenanthrene and 1228 ± 187 µg/kg (n = 5) for pyrene. Desorption experiments were conducted similarly as described in 5.2.1.

The desorption kinetics of phenanthrene and pyrene from the soil is shown in Figure 5-9. The desorbed amounts of the PAHs adjusted to soil weight without the amendment of LMWOA increased from 0 h to 24 h; approximately 6.6% and 6.7% of phenanthrene and pyrene were desorbed within the first 24 h. Desorption of PAH then approached a near steady state after 24 h. A similar kinetic trend was also observed for PAH desorption from soil with the amendment of citric acid. In the presence of 100 mmol/L citric acid, phenanthrene and pyrene desorption from the soil increased during the first 24 h, and then reached desorption equilibration (Figure 5-9). The magnitude of PAHs desorbed from the soil was significantly greater than those desorbed in the controls without the added citric acid ($P<0.001$). At 24 h, 18.1% of phenanthrene and pyrene had been desorbed from soil to the solution containing 100 mmol/L of citric acid. These results indicate that PAH desorption from the soil approached equilibration after about 24 h no matter whether LMWOA was added or not. However, adding LMWOAs to the solution enhanced PAH desorption from the soil significantly ($P<0.001$).

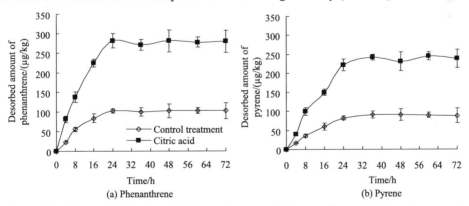

Figure 5-9 Desorption kinetics of (a) phenanthrene and (b) pyrene from soil in the presence and absence of 100 mmol/L citric acid. Desorbed amount is the quantity of PAHs released to solution per unit soil weight. Error bar = ± 1 standard deviation (SD). The P values for the difference between control and citric acid treatments were less than 0.001 for either phenanthrene or pyrene

The extent of PAH desorption from soil after 24 h of contact time in the presence of 0 mmol/L to 1000 mmol/L citric, oxalic and malic acids is shown in Figure 5-10. The amounts of phenanthrene and pyrene desorbed from soil to the solution without LMWOA addition was 102.8 μg/kg and 82.2 μg/kg, respectively, which can be considered to be the baselines, For the solutions containing LMWOAs, the desorbed phenanthrene was up to 395.3 μg/kg, 336.9 μg/kg and 309.9 μg/kg, and desorbed pyrene up to 327.4 μg/kg, 288.8 μg/kg and 266.4 μg/kg when 1000 mmol/L citric, oxalic, and malic acids were added. To quantify this change, desorption ratio (D, %) was estimated as follows

$$D (\%) = C_i / C_o \times 100\%$$

where C_i is the desorbed amount of PAHs per unit weight of soil (μg/kg) at a given LMWOA concentration, and C_o is the concentration of PAH initially present in the contaminated soil (μg/kg).

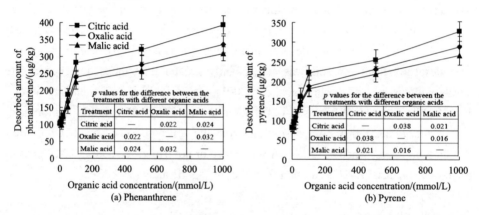

Figure 5-10 Desorption of (a) phenanthrene and (b) pyrene from soil as a function of organic acid concentrations in solution. Error bar = ± 1 standard deviation (SD). The P values for the difference between the treatments with different organic acids (0~1000 mmol/L) were all less than 0.05

The desorption ratios of phenanthrene and pyrene in soil plotted as a function of LMWOA concentration are shown in Figure 5-11. The D values increased with increasing LMWOA concentration; for phenanthrene the D values increased from 6.6% at the baseline to 25.5%, 21.7% and 20.0% in the presence of 1000 mmol/L citric acid, oxalic acid and malic acid, respectively. For pyrene, the D values increased from 5.8% at the baseline to 22.9%, 22.2% and 18.6%, respectively. The D values for phenanthrene were greater than those for pyrene for all three LMWOAs, indicating that phenanthrene was more prone to desorption from the soil.

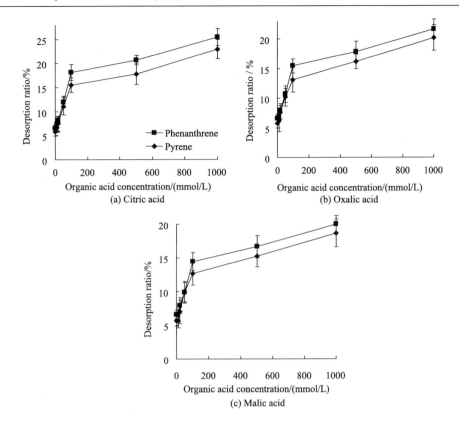

Figure 5-11　Desorption ratios of phenanthrene and pyrene from soil as a function of the concentration of (a) citric acid, (b) oxalic acid, and (c) malic acid in solution. Error bars = 1 standard deviation (SD). The P values for the difference between desorption ratios of phenanthrene and pyrene were less than 0.05

Desorption of phenanthrene and pyrene from the soil varied markedly for the three LMWOAs. Although PAH desorption was enhanced in the presence of all three organic acids, the magnitude of desorption as quantified by the D values followed the descending order of citric acid > oxalic acid > malic acid for both PAHs (Figures 5-10 and 5-11). This is consistent with our previous study that citric acid demonstrated greater efficiency in promoting desorption of spiked PAHs from soils than oxalic acid (Ling et al., 2009). The D value for phenanthrene from soil to 1000 mmol/L citric acid solution was 17.3% and 27.6% greater than that in 1000 mmol/L oxalic acid and malic acid, respectively. The D value for pyrene in 1000 mmol/L citric acid was 13.4% and 22.9% greater, respectively, than the desorption to solutions containing 1000 mmol/L oxalic acid and malic acid.

The increment ratio (r) of phenanthrene and pyrene desorption from soil was calculated according to the following equation

$$r\,(\%) = (C_i - C_{ck}) / C_{ck} \times 100\%$$

where C_{ck} is the desorbed amount of PAH per unit soil weight without the added LMWOA. Larger r values indicate that the increment of PAH desorption by LMWOA becomes more significant.

The calculated r values for phenanthrene and pyrene desorption reached up to 285% and 299%, respectively, when organic acid concentrations increased to 1000 mmol/L. The effectiveness of the three organic acids in promoting desorption followed the order citric > oxalic > malic as indicated by the decreasing r values (Figure 5-12). However, all three LMWOAs were efficient in promoting desorption of phenanthrene and pyrene from soil with r values of >200% (Figure 5-12).

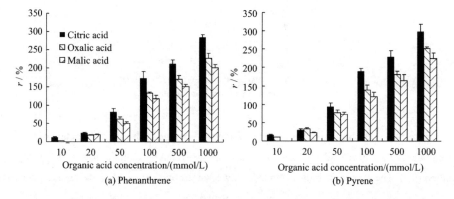

Figure 5-12 Relationship between increment ratio (r) of desorption of phenanthrene and pyrene from soil and organic acid in solution. Error bar = ± 1 standard deviation (SD)

The extent of PAH desorption from soil was in the order of citric > oxalic > malic acid (Figure 5-10, Figure 5-11 and Figure 5-12), the differences result from the different chemical structures and properties of the acids. Citric acid is a ternary acid and could form more stable complexes with metal cations than either oxalic acid or malic acid both of which are binary acids (An et al., 2011). Aliphatic dicarboxylic and tricarboxylic acids generally have concentrations of < 50 mmol/L in soil solution although in some cases they may be up to 650 mmol/L (Hees et al., 1996). In our study, the maximum concentration of LMWOAs is probably greater than their natural concentration in soil solutions. However, even at lower concentrations (10 mmol/L, 20 mmol/L, 50 mmol/L) PAH desorption was enhanced although to a lesser extent.

To further investigate the promotional effects of LMWOAs on PAH desorption, three cycles of desorption were conducted by decanting the supernatant after centrifugation and replenishing the same volume of fresh aqueous solution containing either 0 mmol/L (control) or 1000 mmol/L LMWOAs. Desorption of PAHs from soil was assessed in relation to the first-cycle desorption of PAHs without the added LMWOAs (control) (Figure 5-13). There was a large proportion of the PAHs desorbed to solution during the second and third cycles of desorption. During the second cycle of desorption in citric acid solution, the adjusted desorption value was 0.80 for phenanthrene and 0.78 for pyrene, respectively, which was 46.1% and 19.0% greater than desorption to the control solution without LMWOAs. Similarly, the adjusted desorption value of phenanthrene and pyrene to the solution of citric acid was 0.43 and 0.44, respectively, during the third cycle of desorption, which was equivalent to 1.4 times and 1.3 times desorption in aqueous solution. These results indicate that LMWOAs were still effective in desorbing more PAHs from soil during multiple cycles of the desorption processes, although there was a decreasing magnitude. Thus the first-cycle desorption of phenanthrene and pyrene in citric acid contributed 58.4% and 50.1% of the total PAHs desorbed over all three cycles of desorption. The contributions from the second-cycle and third-cycle desorption were 27.0% and 14.7%, and 31.8% and 18.1% for phenanthrene and pyrene, respectively (Figure 5-13).

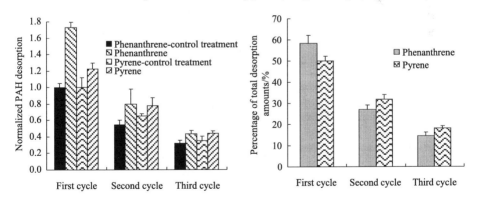

Figure 5-13 Desorption of phenanthrene and pyrene from soil to aqueous organic acid solution during the first-cytle, second-cycle and third-cycle desorption. (a) PAH desorption and (b) percentage of the total desorption. Error bar = ± 1 standard deviation (SD)

5.2.3 Mechanisms of LMWOA-enhanced desorption of PAHs from soils

Here, we take the desorption system as elucidated in 5.2.2 as an example to clarify the mechanisms of LMWOA-enhanced desorption of PAHs from soils. Similar mechanisms were also discussed in "4.2.2, 4.4.3 and 5.1.4".

First, The DOC concentration released from soil was determined and given in Figure 5-14. The DOC concentration increased in the solution after PAH desorption, and was estimated by subtracting the contributions of LMWOAs and PAHs from the total measured DOC. The amount of phenanthrene and pyrene desorbed from soil also increased with increasing DOC (Figure 5-15). The apparent linear relationship between the magnitude of PAH desorbed from the soil was significantly related to the DOC released from the soil ($P < 0.01$).

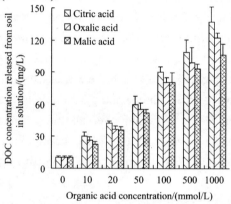

Figure 5-14 Relationship between dissolved organic carbon (DOC) concentration released from soil to solution and the added organic acid. Error bar = ± 1 standard deviation (SD)

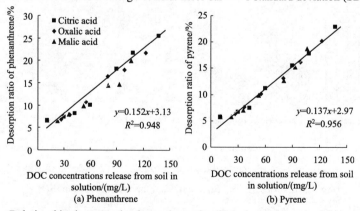

Figure 5-15 Relationships between the desorption ratio of (a) phenanthrene and (b) pyrene from soil with dissolved organic carbon (DOC) concentration released from soil to solution

The measured desorbed amount, desorption ratio and desorption increment ratio of phenanthrene and pyrene from soil in the presence of LMWOAs indicate that the three LMWOAs effectively promoted desorption of PAHs from soil which were linearly related to the increase of DOC in solution. We assume that LMWOAs break the linkages between metal cations and SOM resulting in its release of SOM into solution. Metal cations present in soils can be complexed with functional groups in SOM, and serve as the "bridges" linking them to soil minerals (Saison et al., 2004; Sun et al., 2012). Organic acids are the major active class of substances in root exudates, and they have a strong capability of binding with metal cations in soils (Krishnamurti et al., 1997). Organic matter previously associated with soil minerals via metal-cation bridges could release into solution by the addition of organic acids and become DOC in water. This leads to the decrease of SOM in soils and simultaneous increase of DOC in solution, both favoring PAH desorption from soils (Sun et al., 2012). Furthermore, plant root-excreted organic acids could facilitate the dissolution of soil minerals (Drever and Stillings, 1997), which also favors the release of SOM to solution.

Soil organic matter has been viewed as the predominant site for hydrophobic organic compound sorptions (Chiou et al., 1979; Weber and Huang, 2003). Soils containing larger SOM contents generally demonstrate a strong uptake of organic compounds from water. In our study, the increased DOC in aqueous solution resulting from the release of SOM by LMWOAs reduced SOM content, and correspondingly, increased PAH desorption. The reduction of SOM content (R_{SOM}, mg/g) in the soil was estimated as

$$R_{SOM} = (C_{DOC\text{-}i} - C_{DOC\text{-}o}) \times V / (M_{soil} \times 1000)$$

where $C_{DOC\text{-}i}$ and $C_{DOC\text{-}o}$ are DOC concentrations (mg/L) in solution after desorption of PAHs from soil in the presence and absence of added LMWOAs; M_{soil} is soil mass (0.50 g), and V is aqueous solution volume (25 mL). Assuming that SOM released by LMWOA has the same capacity of retaining PAHs as the original SOM, and that in solution has little sorption capacity for PAHs, then a predicted increment ratio ($r_{predicted}$, %) of PAH desorption from soil in the presence of LMWOAs can be estimated by

$$r_{predicted} = Q_{R\text{-}SOM} \times 100 / M_{total}$$

where M_{total} is the initial concentration of PAH in soil (mg/kg), and $Q_{R\text{-}SOM}$ is the reduced sorption of PAHs (mg/kg) caused by the loss of SOM in the presence of LMWOAs. $Q_{R\text{-}SOM}$ can be calculated as

$$Q_{R\text{-}SOM} = Q_{SOM} \times R_{SOM} / f_{oc}$$

where Q_{SOM} is the sorption of PAHs in soil (mg/kg), f_{oc} is the soil organic carbon content (mg/g), and R_{SOM} is the reduced amount of soil organic carbon (mg/g). Q_{SOM} can be obtained from

$$Q_{SOM} = (1 - D_o) \times M_{total}$$

where D_o is the desorption ratio of PAH from soil without the added LMWOA in solution. Therefore, by combining the above equations, $r_{predicted}$ could be estimated as

$$r_{predicted} = (1 - D_o) \times (C_{DOC\text{-}i} - C_{DOC\text{-}o}) \times V / (M_{soil} \times f_{oc} \times 10)$$

The estimated $r_{predicted}$ values are presented in Table 5-6. Clearly, the $r_{predicted}$ values were dramatically less than the measured values ($r_{measured}$) indicating that, in addition to the release of SOM into solution, other mechanisms also contribute to LMWOA-enhanced PAH desorption from soil.

Table 5-6 Comparison of predicted ($r_{predicted}$, %) and measured increment ratio ($r_{measured}$, %) of desorption of phenanthrene and pyrene from soil to the solution with added organic acids

Organic acid	C_{OA}[a] /(mmol/L)	C_{DOC}[b] /(mg/L)	$r_{measured}$ /%[c] Phenanthrene	Pyrene	$r_{predicted}$ /%[d] Phenanthrene	Pyrene
Ctric acid	10	30.2	12.2	16.6	6.7	6.8
	20	42.5	24.0	30.6	11.0	11.1
	50	59.9	51.6	94.0	16.9	17.1
	100	90.1	174.2	169.9	27.3	27.5
	500	108.6	227.7	209.4	33.6	34.0
	1000	137.3	284.7	298.6	43.5	43.9
Oxalic acid	10	26.7	2.9	9.9	5.5	5.6
	20	37.0	17.7	32.5	9.1	9.1
	50	55.5	61.6	77.0	15.4	15.6
	100	80.2	148.1	128.3	23.9	24.1
	500	99.2	169.9	181.7	30.4	30.7
	1000	122.4	227.9	251.6	38.4	38.7
Malic acid	10	22.6	—	—	4.1	4.2
	20	35.8	19.4	21.9	8.7	8.7
	50	52.4	49.6	72.0	14.4	14.5
	100	80.6	118.3	120.2	24.0	24.3
	500	93.3	121.3	165.1	28.4	28.7
	1000	106.1	201.6	224.3	32.8	33.1

Note: a C_{OA} is the added concentrations of organic acid in solution; b C_{DOC} is the measured DOC concentration in solution; c,d $r_{predicted}$ and $r_{measured}$ are predicted and measured increment ratio of desorption of phenanthrene and pyrene from soil.

The significantly positive relationship between PAH desorption and DOC in solution indicates the important role of DOC (Figure 5-15) in interacting strongly with hydrophobic organic compounds, and changing their fate and transport in the environment (Maxin and Kögel-Knabner, 1995). The association of DOC with PAHs has been identified as an important process which increases the solubility of PAHs in water (Chiou et al., 1986; Haftka et al., 2010). The corresponding partition coefficient (K_{doc}) of phenanthrene with DOC was 8318 L/kg (Mott, 2002), which was approximately twice that measured between soil and water (Ling et al., 2009; Gao et al., 2010). DOC may have a larger potential to sorb PAHs than bulk soils. Therefore, increased DOC concentration in solution could result in more desorbed PAHs from soil.

5.3 Impact of LMWOAs on the availability of PAHs in soil

Understanding of PAH availability in the soil environment will have considerable benefits for risk assessment purposes and remediation strategies for contaminated soils. This section was designed to examine the impacts of two LMWOAs (citric and oxalic acid) on PAH availability in contaminated soils (Ling et al., 2009). Phenanthrene and pyrene were used as representative PAHs. The method involved non-exhaustive extraction with *n*-butanol, and PAHs by butanol-extraction was proven to be significantly bioavailable to organisms (Liste and Alexander, 2002).

Two zonal soils (Paleudult and TypicPaleudalfs), previously free of PAHs, were collected from the A (0~20 cm) horizon in Wuhan and Nanjing provinces in China. The Paleudult soil had a pH of 4.74, 9.47 g/kg soil organic carbon content (f_{oc}), and consisted of 39.2% clay, 51.6% sand, and 9.20% silt. The TypicPaleudalfs soil had a pH of 6.02, 14.3 g/kg soil organic carbon content (f_{oc}), and consisted of 24.7% clay, 61.9% sand, and 13.4% silt. Soil were air-dried, sieved, and spiked with PAH in acetone (Gao et al., 2009). Availability test was conducted using microcosm experiments. PAH-spiked soils were packed into amber glass microcosms similar to those reported in the literature (Macleod and Semple, 2003). Citric acid or oxalic acid solution (0~1000 mmol/L) was added. In some experiments, a 0.2% NaN_3 solution, which was proven to be effective to inhibit the microbial growth in soil (Folberth et al., 2009), was added as a microbe-sterilizing treatment. Soil water content was adjusted to 30% of the soil water-holding capacity. After incubation at 25℃ for 0 day, 15 day, 30 days, 45 days or 60 days, soils were sampled, PAHs were extracted by *n*-butanol solution, and their concentrations determined.

5.3.1 Impact of LMWOAs on the butanol-extractable PAHs in soils

Here the method involved non-exhaustive extraction with *n*-butanol was selected to assess the availability of PAHs in soils. Figure 5-16 illustrates that the extractable amounts of phenanthrene and pyrene in test two soils increased as a function of citric and oxalic acid concentrations from 0~57.6 g/kg and 0~27.0 g/kg, respectively. LMWOA addition significantly increased the extractable amount of both phenanthrene and pyrene in a concentration-dependent manner, and the higher concentration of added LMWOA resulted in the larger amounts of extractable PAHs in soils. For example, the concentrations of extractable phenanthrene and pyrene in Paleudult significantly increased from 11.51 mg/kg and 10.75 mg/kg to 22.13 mg/kg and 16.16 mg/kg with the increase of citric acid concentration from 0 g/kg to 57.6 g/kg, respectively. Similarly, the concentrations of extractable phenanthrene and pyrene increased 68.8% and 25.7% with oxalic acid concentration from 0 g/kg to 27.0 g/kg. These findings indicate that LMWOAs significantly enhance the availability of the tested PAHs in these soils. The same trend was observed for TypicPaleudalfs treatments. In addition, the degree of extractable PAHs differed with different organic acids, although an obvious increase was observed with both LMWOAs. The extractable amounts of phenanthrene and pyrene with citric acid were generally much higher than those with oxalic acid (Figure 5-16).

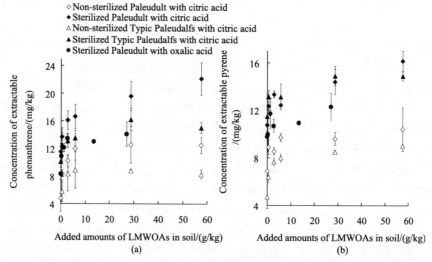

Figure 5-16 Concentrations of extractable phenanthrene (a) and pyrene (b) as a function of added amounts of LMWOAs in tested soils after 60 days incubation. Error bars are standard deviations (SD)

Soil organic matter (SOM) strongly influences the availability of PAHs (Hatzinger and Alexander, 1995; Nam et al., 1998). Previous work has indicated that high soil organic carbon (SOC) content is associated with a more potent inhibitory effect on PAH extractability (Gao et al., 2009). Our findings were consistent with this. The SOC contents of tested two soils Paleudult and TypicPaleudalfs were 9.47 g/kg and 14.3 g/kg. As seen in Figure 5-16, although the addition of LMWOAs significantly enhanced the extractable amounts of phenanthrene and pyrene in both soils, the amounts of extractable PAHs in TypicPaleudalfs were generally lower than those in Paleudult with a relatively lower SOC content (9.48 g/kg), irrespective of the addition of organic acids, and this trend was more significant for phenanthrene. These results meant that the higher SOC led to the lower extractability of PAHs in soil, and the organic matter content of the soils had a large influence on the availability of PAHs.

In addition, more significant increment of extractable PAHs was observed with citric acid in Paleudult than in Typic Paleudalfs. As shown in Figure 5-16, the slope of the curve of the extractable concentration of PAHs as a function of added amount of organic acid in the former soil was generally bigger than the latter. This indicates that lower the SOC of soils, higher the degree of the increment of extractable PAHs by LMWOAs. For instance, the concentrations of extractable phenanthrene in non-sterilized Paleudult and TypicPaleudalfs increased by 161.8% and 125.3%, and in sterilized these two soils increased by 92.3% and 48.7% with 57.6 g/kg citric acid, respectively.

Figure 5-17 illustrated the increment rate (r) of extractable phenanthrene and pyrene by the addition of citric acid and oxalic acid in sterilized Paleudult aged for 60 days. r was calculated as follows

$$r = (C_i - C_o) / C_o \times 100\%$$

where, C_i and C_o were the extractable PAH concentration in soil with and without LMWOA addition, respectively. Higher the r value indicates the more significant promotion of the PAH extractability by organic acid. Clearly, the promotion of the extractability of phenanthrene in soils by these two organic acids were more significant than that of pyrene, as displayed by the much higher r values of the former than the latter. This may owe to the higher solubility in water, lower K_{ow}, and smaller molecular weight of phenanthrene.

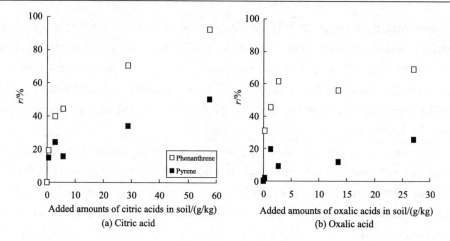

Figure 5-17　The increment rate (r) of extractable phenanthrene and pyrene by citric (a) and oxalic acid (b) in sterilized Paleudult aged for 60 days. r was assessed as $r=(C_i-C_o)/C_o$, where C_i is the extractable concentration in soil with a certain concentration of organic acid, and C_o is extractable concentration in soil without organic acid addition

One notes that the availability of phenanthrene in soils diminished more rapidly, and in the same soil, the dissipation ratio of available phenanthrene was clearly higher than that of pyrene. As shown in Figure 5-18, 64.3% and 80.4% of extractable phenanthene degraded in sterilized and non-sterilized Paleudult without acid, respectively, whereas the number was 44.7% and 69.3% for pyrene in these soils.

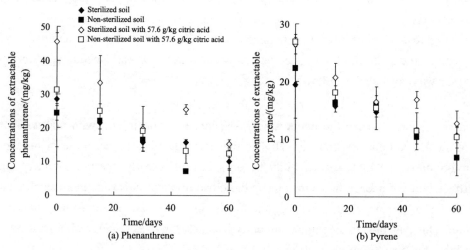

Figure 5-18　Concentrations of extractable phenantherene (a) and pyrene (b) in sterilized and non-sterilized Paleudult as a function of aging time. Error bars are standard deviations (SD)

Similarly, 66.3% and 52.9% of extractable phenanthrene and pyrene disappeared in sterilized Paleudult with 57.6 g/kg citric acid. These results indicate that PAHs with higher molecular weight and more benzene rings were more recalcitrant in soils (Sabate et al., 2006). The rapid establishment of a potent phenanthrene-degrading microbial consortium in the soil was observed in previous studies (Semple et al., 2003). However, because of the higher molecular weight and greater number of benzene rings, pyrene was more recalcitrant to degradation; thus, contaminated soils may require a longer period to establish an effective microbial consortium.

Extractability with mild extractants diminishes over time (Hatzinger and Alexander, 1995; Chung and Alexander, 1998). For this reason, the following experiment was designed to determine the quantity of PAHs that could be extracted with butanol at various intervals. Figure 5-18 shows that the concentrations of extractable phenanthrene and pyrene in the tested soils decreased as a function of incubation time, from 0 day to 60 days. In freshly-spiked, non-sterilized soil, 24.36 mg/kg~31.19 mg/kg phenanthrene and 22.34 mg/kg ~26.83 mg/kg pyrene were recovered. PAH extraction was highest in the presence of 57.6 g/kg citric acid. In contrast, only 4.774 mg/kg~12.49 mg/kg phenanthrene and 6.873 mg/kg~10.40 mg/kg pyrene were recovered after aging for 60 days. A similar tendency was observed in other treatments. This suggests that the aging process significantly affects the availability of PAHs in soils. These PAHs were more readily available at the start of incubation, but their availabilities decreased rapidly with increased soil-PAH contact time. These results may be explained by the slow molecular diffusion of contaminants into micropores within sorbent phases (organic matter), as described in literatures (Pignatello and Xing, 1996).

Decreased PAH extractability over a 60-day period may provide a mean of predicting the decline in bioavailability to microorganisms as a result of aging. As shown in Figure 5-18, the dissipation of available PAH fraction from non-sterilized soils was primarily due to intrinsic microbial degradation. After 60 days of incubation, the extractable concentrations of phenanthrene and pyrene were 10.18 mg/kg and 10.75 mg/kg in sterilized soils with no organic acid, but only 4.774 mg/kg and 6.873 mg/kg in non-sterilized soils (Figure 5-18). Similar results were also observed in other treatments. Compared to microbe-inhibited soils, the overwhelmingly larger dissipation of extractable residues of PAHs in non-sterilized soils indicates that microbial biodegradation was the major contributor to dissipation of the available PAH fraction.

Microbial catabolism has been reported in many studies, and has been described as a principal mechanism for the removal of available organic contaminants, such as PAHs, from soils (Semple et al., 2003). In sterilized soils, the concentrations of extractable PAHs were greater in 60-day-aged soils compared to non-sterilized soils. The decrease in extractable PAH over 60 days was 80.4% for phenanthrene and 69.3% for pyrene in non-sterilized soils; these values were higher than those in sterilized control soils (64.3% and 44.7%, respectively). The 60 days decrease in extractable PAH is attributable to microbial biodegradation of the available fraction and its transformation into other forms in the soils. In sterilized soils (microbial activity inhibited), the decrease in PAH availability can predominantly be ascribed to non-microbial transformation into other forms. These findings further demonstrate the dominant role of microbial biodegradation in desorbing PAH fractions in natural soils. Nevertheless, the addition of 57.6 g/kg citric acid increased the extractable fraction of both phenanthrene and pyrene in both sterilized and non-sterilized soils throughout the sampling period.

5.3.2 Mechanism discussions

As known, the soil organic matter was the predominant pool for hydrophobic organic chemicals in soil (Chiou et al., 1979, 1998). Previous work has indicated that high soil organic carbon (SOC) content is associated with a more potent inhibitory effect on PAH extractability (Ling et al., 2009). This was further proven by the lower PAH extractability in Typic Paleudalfs with higher SOC content in this work. Metal cations in soil act as a "bridge" in organic matter-soil mineral complex (Gao et al., 2003, 2006). LMWOAs found in the environment comprise monocarboxylic, dicarboxylic, and tricarboxylic acids including compounds containing unsaturated carbon and hydroxyl groups. The LMWOA function as ligands can bind metal cations, and thus the organic matter complexing the soil mineral through the "bridge" of metal cations releases into solution (Lu et al., 2007), resulting in the decreased soil organic matter solid and increased dissolved organic matter (DOM) by the addition of LMWOA (Gao et al., 2012). It can be speculated that the increased DOMs were derived from the release of organic matter-soil mineral complex as a result of dissolved metal cations. As such, the enhanced PAH availability in soil by organic acids could be dominantly attributed to the release of soil organic matter into solution phase.

As stated previously, microbial biodegradation was the major contributor to dissipation of the available PAH in soil. However, in sterilized soil, the decrease of extractable PAH fraction with aging time was obvious (Figure 5-18). The forms of nonionic organic chemicals (NOCs) in soil environment have been reported in literatures (Lesan and Bhandari, 2004), and can be fractionated into two fractions including extractable fraction and bound residue. In theory, the formation of bound residue would reduce the availability of NOCs in soil. The transformation of PAHs into bound residue in 0~16 week was also observed in our previous work (Gao et al., 2009). Thus the decrease of butanol-extractable PAH in tested soils can be predominantly ascribed to transformation to its bound residual fraction.

Significant amounts of organic acids released from plant roots may affect PAH elimination by activation of abiotic oxidation of these chemicals. As documented, hydrogen peroxide is elicited as plant-stress responses (Muratova et al., 2009). In the presence of hydrogen peroxide, carboxylic acids produce peroxy-acids, the breakup of which leads to the release of free hydroxyl radicals (OH^-). These strong nonspecific oxidants can rapidly degrade a variety of organic chemicals including PAHs, and organic acids act as catalysts in this process (N'Guessan et al., 2006). However, in this work, the overwhelming increase of available PAHs by addition of organic acids indicates that the oxidation process involving organic acids is minor.

One notes that in general, soil solution concentrations of aliphatic dicarboxylic and tricarboxylic acids are below 50 μmol/L, and in some cases concentrations up to 650 μmol/L have been reported (Hees et al., 1996). In this work, to obtain more convinced results the added amounts of citric or oxalic acid were generally larger than their natural concentrations in soils. On the other hand, the clearly positive effects of organic acid on availabilities of PAHs provide a new method to enhance the remediation efficiency by LMWOA addition.

5.4 Elution of soil PAHs using LMWOAs

The elution characteristics of PAHs in soils are well known to affect their mobility in the environment and thus affect their transformation and fate. Column elution methods can be used to overcome some of the drawbacks with batch equilibrium methods, and they have also been regarded to be more realistic in simulating the leaching processes that occur in the field (Jackson et al., 1984). Column elution experiments enable investigators to investigate the dynamic release characteristics and availability variations of PAHs in soils at variable LMWOA concentrations. In addition, the elution characteristics of PAHs in soils are important indexes to assess their environmental behaviors. The elution characteristics determine the mobility of PAHs in the environment, and affect their transformation and ultimate fates. An additional advantage of column elution methods is that they allow eluates to be collected for chemical analysis, e.g., the metal ions and PAHs.

In this section, we seek to examine the impacts of two LMWOAs (citric acid and malic acid) on the elution of phenanthrene and pyrene as representative PAHs in contaminated soils using soil column experiments.

Brown-red soil, yellow-brown soil and red soil, all previously free of PAHs, were collected from the A horizon (0~20 cm). They are the typical zonal soils of China. Their basic properties are listed in Table 4-2. PAH-spiked soil was obtained according to Kong et al. (2013). Phenanthrene and pyrene dissolved in acetone were added to the sterilized soils. After the acetone evaporated, soil water content was adjusted to 30% of the soil water-holding capacity. We incubated the PAH-spiked soils for 30 days at 25℃ to obtain homogenized soil samples. The final concentrations of phenanthrene and pyrene in spiked yellow-brown soil, brown-red soil, and red soil were 95.53 mg/kg and 82.39 mg/kg, 94.73 mg/kg and 82.72 mg/kg, and 93.99 mg/kg and 81.90 mg/kg of dry soil, respectively. Within these spiked soils, the n-butanol-extractable fractions of phenanthrene and pyrene were 36.50 mg/kg and 38.29 mg/kg, 37.01 mg/kg and 37.88 mg/kg, and 36.56 mg/kg and 37.87 mg/kg of dry soil, respectively.

In order to simulate the actual state of PAH contamination in rhizosphere soil, we designed the dimensions of the soil column (Kong et al., 2013). The experimental setup consisted of an elution column with a filter funnel and flow controller. The elution column was a glass tube with a 50 mm inner diameter. The net height of the elution column from the top to the filter funnel was 150 mm. Soil samples were placed into the elution column with a soil loading height of 100 mm. A quartz sand layer of 10 mm thickness was placed on top of the soil layer to distribute the elution solution well and avoid the impingement effect of droplets on the soil surface layer. In addition, the filter funnel at the bottom of elution column performed the function of filtering the elution solution. The inner diameter/net height ratio was 1 : 3, which can ensure the flow uniformity of LMWOA solution in soil column. Citric or malic acid solutions were fed into the elution column uniformly via a flow controller. The eluates were collected in a conical flask. For each experiment, after every 30 mL of eluate collected, the conical flask was replaced. In all, the eluates were collected 16 times with a total volume of 480 mL. When elution was complete, we separated averagely the soil column into top, middle and bottom layers, and the depth of the top, middle and bottom layers was 0~33 mm, 33~66 mm and 66-100 mm, respectively. The soils in each layer were then sampled to determine the total and available PAHs. Basing on the PAH concentrations in the soil layers with different heights, the dynamic variations of PAHs in soil column during elution by LMWOAs could be investigated.

5.4.1 Elution of PAHs in soil columns by LMWOAs

Based on the original and residual concentrations of PAHs in soils, we calculated the percentage PAH elution by LMWOAs, using the following equation

$$R_e = \frac{C_0 - C_e}{C_0} \times 100\%$$

where R_e is the percentage PAH elution by LMWOAs, and C_0 and C_e are the original and residual concentrations, respectively, of PAHs in soil (mg/kg). The mobilized amounts and elution percentages of phenanthrene and pyrene in a yellow-brown soil column by LMWOAs are listed in Table 5-7. As seen, with increasing concentrations of LMWOAs, the mobilized amounts and elution percentages of both phenanthrene and pyrene also increased. When the concentration of citric acid was 10 mmol/L, the mobilized amounts and elution percentages were 20.55 mg/kg (21.51%) and 31.09 mg/kg (37.73%) for phenanthrene and pyrene, respectively. Compared to the control, these increases were 5.85 mg/kg (6.12%) and 15.98 mg/kg (19.39%) for phenanthrene

Table 5-7 Elution of polycyclic aromatic hydrocarbons in the yellow-brown soil column by low-molecular-weight organic acids

Organic acid		Phenanthrene		Pyrene	
Type	Concentration /(mmol/L)	Mobilized amount /(mg/kg)	Elution/%	Mobilized amount/(mg/kg)	Elution/%
Control	0	14.70±1.50	15.39±1.57	15.11±1.31	18.34±1.59
Citric acid	10	20.55±4.27	21.51±4.47	31.09±0.46	37.73±0.55
	40	54.38±1.43	56.93±1.49	36.28±0.76	44.03±0.92
	80	59.09±0.64	61.85±0.67	59.13±2.31	71.77±2.81
Malic acid	10	31.21±5.86	32.67±6.14	32.27±7.64	39.16±9.28
	40	58.42±1.77	61.15±1.85	38.36±6.20	46.56±7.52
	80	59.52±2.64	62.31±2.76	55.90±6.45	67.84±7.83

and pyrene, respectively. However, at a concentration of 80 mmol/L, the mobilized amounts and elution percentages were higher at 59.09 mg/kg (61.85%) and 59.13 mg/kg (71.77%) for phenanthrene and pyrene, respectively. The elution percentages for both phenanthrene and pyrene were more than three times higher than those of the control. Similar results were also observed for malic acid. When the concentration of malic acid was 80 mmol/L, the mobilized amounts and elution percentages reached 59.52 mg/kg (62.31%) and 55.90 mg/kg (67.84%) for phenanthrene and pyrene, respectively. Both citric acid and malic acid were shown to elute large amounts of phenanthrene and pyrene from the soil column compared to the control.

5.4.2 Distributions of PAHs in soil columns

The PAH concentrations in the soil layers at different heights indicated the dynamic variations of PAHs in the soil column during elution by LMWOAs. For yellow-brown soil, we measured the PAH concentrations in the soil layers at different heights. After elution, the soil column was partitioned into top, middle, and bottom layers and the soils in each layer were sampled. The total PAHs in the soil samples were then determined. The distributions of phenanthrene and pyrene in the yellow-brown soil column following citric and malic acid elution are shown in Figure 5-19. The residual concentrations of phenanthrene and pyrene in each soil layer decreased during the elution by citric or malic acid, which indicated that LMWOAs promoted the elution of PAHs from the yellow-brown soil. Moreover, the distributions of PAHs varied for different concentrations of LMWOAs. When the concentration of LMWOAs was equal to zero (control), the phenanthrene concentration in soil layers followed the order top layer > middle layer > bottom layer, and the pyrene concentration followed the order top layer > middle layer ≈ bottom layer. However, when the concentration of LMWOAs was greater than zero (5~80 mmol/L), the distributions of PAHs differed substantially from those of the control. When the concentrations of LMWOAs were low (5 mmol/L and 10 mmol/L), the PAH concentrations in the soil layers followed the order bottom layer > middle layer > top layer. With an increase in the concentrations of LMWOAs, this trend gradually weakened, until at the concentration of 80 mmol/L, the PAH concentrations in all soil layers were almost equal.

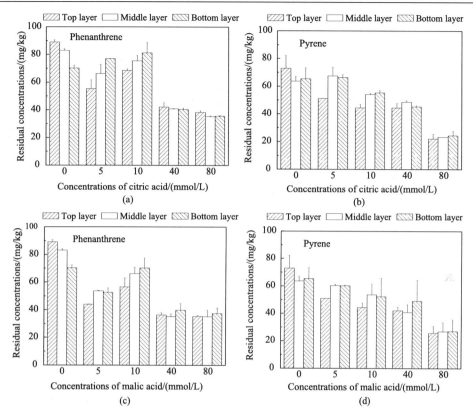

Figure 5-19 The distributions of phenanthrene and pyrene in the yellow-brown soil column via citric acid and malic acid elution (Error bars indicate standard deviations)

5.4.3 Butanol-extractable and nonextractable PAHs in soil columns

The residual concentrations of *n*-butanol-extractable and nonextractable phenanthrene and pyrene by *n*-butanol in the yellow-brown soil column by citric and malic acid elution are shown in Figure 5-20, which shows that both the residual concentrations of *n*-butanol-extractable and nonextractable PAH generally decreased with increasing organic acid concentration. The original concentrations of nonextractable phenanthrene and pyrene in the yellow-brown soil were 59.03 mg/kg and 44.10 mg/kg, respectively. At the organic acid concentration of 0 mg/kg (control), the residual concentration of nonextractable phenanthrene remained almost unchanged, and the residual concentration of non-available pyrene decreased to 28.35 mg/kg. When the concentration of citric acid was 10 mmol/L, the residual concentrations of nonextractable phenanthrene and pyrene were 56.17 mg/kg and 24.29 mg/kg, respectively. However, when the concentration of citric acid increased to 80 mmol/L, their residual concentrations decreased to 21.74 mg/kg and 11.74 mg/kg, respectively, and the decreases in nonextractable phenanthrene and pyrene relative to their initial amounts were 63.17% and 73.38%, respectively. A similar result was obtained when malic acid was used, as shown in Figure 5-20. At the malic acid concentration of 80 mmol/L, the residual concentrations of nonextractable phenanthrene and pyrene were 16.26 mg/kg and 15.79 mg/kg, respectively, and the decreases in nonextractable phenanthrene and pyrene relative to their initial amounts were 72.45% and 64.20%, respectively. The results indicated that after elution by LMWOAs, some initially nonextractable PAHs in soils could be transformed into n-butanol-extractable PAHs.

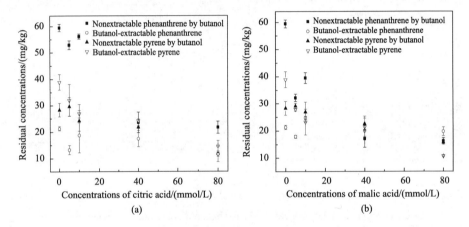

Figure 5-20 The residual concentrations of n-butanol-extractable and nonextractable phenanthrene and pyrene by n-butanol in the yellow-brown soil column via citric (a) and malic acid (b) elution
(Error bars indicate standard deviations)

5.4.4 Impact of soil type on PAH elution

The elution percentages and residual concentrations of PAHs in three different soil columns by citric acid are presented in Table 5-8, which shows that after elution by citric acid, the total and nonextractable residual concentrations of PAHs decreased significantly. Moreover, with an increase in citric acid concentration, a trend was detected for the elution percentages of the total PAHs in three different soil columns to increase and nonextractable residual concentrations of PAHs to decrease. When the concentration of citric acid was 80 mmol/L, the elution percentages of the total

Table 5-8 Elution percentages and residual concentrations of PAHs in three different soil columns by citric acid

PAHs	Citric acid concentration /(mmol/L)	Yellow-brown soil		
		Elution percentage/%	C_A/(mg/kg)	C_{NA}/(mg/kg)
Phenanthrene	5	30.69±0.57	13.28±1.71	52.92±1.65
	10	21.51±4.47	18.81±6.60	56.17±0.98
	40	56.93±1.49	17.50±3.03	23.65±3.89
	80	61.85±0.67	14.70±1.99	21.74±2.31
Pyrene	5	25.14±3.45	36.05±6.05	29.63±3.52
	10	39.46±4.41	29.48±0.68	20.39±4.14
	40	44.03±0.92	24.26±0.69	21.85±0.46
	80	71.77±2.81	11.52±0.71	11.74±3.02
PAHs	Citric acid concentration /(mmol/L)	Brown-red soil		
		Elution percentage/%	C_A/(mg/kg)	C_{NA}/(mg/kg)
Phenanthrene	5	25.95±2.40	26.59±2.50	43.56±4.71
	10	40.55±4.14	28.65±2.61	27.67±6.03
	40	50.72±1.70	25.54±3.46	21.15±4.30
	80	56.14±1.89	20.33±1.70	21.21±2.70
Pyrene	5	20.11±2.91	37.57±1.94	28.52±1.02
	10	23.59±0.35	30.10±2.37	33.10±2.42
	40	29.98±1.95	27.22±0.62	30.68±1.93
	80	44.44±2.15	23.04±1.97	22.91±3.56
PAHs	Citric acid concentration /(mmol/L)	Red soil		
		Elution percentage/%	C_A/(mg/kg)	C_{NA}/(mg/kg)
Phenanthrene	5	23.66±1.12	36.62±0.81	35.13±1.29
	10	26.03±1.23	33.07±0.11	36.45±2.16
	40	46.94±1.20	26.70±0.80	23.18±1.15
	80	46.15±2.57	26.72±0.65	23.89±2.12
Pyrene	5	6.63±1.73	40.87±0.49	35.60±1.29
	10	17.24±1.55	36.09±1.29	31.69±1.80
	40	67.25±5.44	7.89±0.33	18.94±3.27
	80	72.03±1.76	9.85±0.02	13.06±1.45

Note: C_A and C_{NA} denote the residual concentrations of n-butanol-extractable and nonextractable PAHs by n-butanol in soil.

phenanthrene and pyrene in yellow-brown soil, brown-red soil, and red soil were 61.85% and 71.77%, 56.14% and 44.44%, and 46.15% and 72.03%, respectively. The corresponding mobilized amounts were 59.09 mg/kg and 59.13 mg/kg, 53.18 mg/kg and 36.76 mg/kg, and 43.38 mg/kg and 58.99 mg/kg. When the concentration of citric acid increased from 5 mmol/L to 80 mmol/L, the nonextractable residual concentrations of phenanthrene in yellow-brown soil, brown-red soil, and red soil decreased from 52.92 mg/kg, 43.56 mg/kg, and 35.13 mg/kg to 21.74 mg/kg, 21.21 mg/kg, and 23.89 mg/kg, respectively. When the concentration of citric acid was 80 mmol/L, the amounts of nonextractable phenanthrene in yellow-brown soil, brown-red soil, and red soil were 37.29 mg/kg, 36.51 mg/kg, and 33.54 mg/kg, respectively. Therefore, after elution by citric acid with a concentration of 80 mmol/L, 63.17%, 63.25%, and 58.40% of nonextractable phenanthrene in yellow-brown soil, brown-red soil, and red soil, respectively, were transformed into n-butanol-extractable phenanthrene. The elution percentages of PAHs by citric acid differed depending on the soil type. For red soil, when the concentration of citric acid was low (5~20 mmol/L), the elution percentage of the total pyrene was relatively low with a value below 20%. However, at a citric acid concentration of 80 mmol/L, the elution percentage of the total pyrene reached 72.03%, which was close to that in yellow-brown soil and far larger than that in brown-red soil. In addition, for all three soils when the concentration of citric acid was 5 mmol/L, the nonextractable residual concentration of pyrene was lowest for brown-red soil. However, when the concentration of citric acid was increased to 80 mmol/L, the nonextractable residual concentration of pyrene in brown-red soil was 22.91 mg/kg, which was almost twice that of yellow-brown soil or red soil.

5.4.5 Mechanisms of LMWOA-enhanced elution of soil PAHs

The increase in elution percentages and availabilities was primarily attributable to the activation of soil organic matter (SOM) by LMWOAs. Pignatello and Xing (1996) confirmed that PAHs in soils could combine with SOM to form bound PAHs. They also become positioned into the micropores in soil aggregates through intraparticle diffusion, and with time, they can enter the deeper sorption sites and become bound. Both these mechanisms result in a decrease in availability of soil PAHs. A study by Nardi et al. (1997) also revealed that organic pollutants containing PAHs in soils were mainly bound in the SOM and greatly affected by the content and configuration of SOM. Raber et al. (1998) found that PAH desorption increased linearly with an increasing concentration of dissolved organic matter (DOM) in soil solution. Thus, we can assume that LMWOAs can dissolve a portion of SOM into solution during the elution process. To validate this assumption, we determined the DOM in the soil solution after the addition of citric acid. When the concentration of citric acid was 100 mmol/L, the equilibrium concentration of DOM in solution was 90.13 mg/L. Compared to the control (the concentration of citric acid was zero), the increase in DOM concentration was 79.56 mg/L. This finding demonstrated that LMWOAs can dissolve a portion of SOM into solution during the elution process, resulting in an increase in the elution percentages and availability of PAHs.

The SOC content of the three soils followed the order yellow-brown soil > brown-red soil > red soil. Our previous study using batch equilibrium experiments indicated that a higher SOC content led to a lower extractability of PAHs in soil (Ling et al., 2009). Thus, according to this previous batch equilibrium experimental study, after elution by LMWOAs, the residual nonextractable PAHs in soils should follow the order yellow-brown soil > brown-red soil > red soil. However, the results of this study using soil column elution experiments did not follow this order (Table 5-8). For example, at the citric acid concentration of 5 mmol/L, the residual amounts of nonextractable phenanthrene in yellow-brown soil, brown-red soil, and red soil were 52.92 mg/kg, 43.56 mg/kg, and 35.13 mg/kg, respectively. This order was consistent with that in previous studies. However, when the citric acid concentration was 80 mmol/L, the residual amounts of nonextractable phenanthrene in yellow-brown soil, brown-red soil, and red soil were 21.74 mg/kg, 21.21 mg/kg, and 23.89 mg/kg, respectively. The difference between residual nonextractable phenanthrene concentrations in the three soils was minor. These results indicated that other than SOM, other physical chemistry properties of the soils and soil column elution properties may affect the release and availability of PAHs. For example, the soil particle compositions of the three soils were different, so the porosities of soil columns would also vary. Despite an identical volume of elution solution being used, the actual flow conditions were different, hence affecting the diffusion of compounds in solution containing LMWOAs, DOM, and PAHs, finally affecting the release of PAHs from soils to the liquid phase.

5.4.6 Relationship between the elution of PAHs and the dissolution of metal ions

The organic matter in soils always combines with minerals to form an organic-mineral complex (Li et al., 2006). Thus, the sorption and immobilization of

PAHs in soils results from the combined actions of SOM and soil minerals (Huang et al., 1996). Metal ions (e.g., Fe, Ca, and Al) have an important role in the formation and stabilization of the organic-mineral complex. Yang et al. (2001) found that between 52% and 98% of SOM was bound in to an organic metal ion-mineral complex. Because organic pollutants in soils can be sorbed or bound by SOM, organic pollutants can be fixed in the organic metal ion-mineral complex. Moreover, Yang et al. (2001) also found that the desorption of PAHs had a relationship with the dissolution of metal ions from soils. To analyze the mechanisms of enhanced release and availability of PAHs in the soil columns by LMWOAs, we determined the concentrations of Fe, Al, Ca, Mg, and Mn ions in the eluates.

The average concentrations of metal ions (Fe, Al, Ca, Mg, and Mn) in eluates from the yellow-brown soil column at different concentrations of LMWOAs are presented in Table 5-9. Compared to the control, the average concentrations of all metal ions were far larger, and all the metal ion concentrations in the eluates increased with increasing concentrations of LMWOAs. Moreover, to analyze the dynamic process of the release of PAHs and metal ions from the soil column, we chose an elution experiment (yellow-brown soil; citric acid concentration 40 mmol/L) to determine the concentrations of metal ions (Fe, Al, Ca, Mg, and Mn) in each 30 mL of eluate. The cumulative, volume-dependent patterns of the concentration curves for Fe, Al, Ca, Mg, and Mn ions in the eluates are shown in Figure 5-21. The patterns of the concentration curves for the five different metal ions were similar. When the cumulative volume was less than 120 mL, with the increase in the cumulative volume of eluates, all metal ion concentrations increased; at the cumulative volume range of 120~360 mL, all the metal ion concentrations slightly fluctuated, either up or down; and when the cumulative volume was larger than 360 mL, all the metal ion concentrations decreased with increasing cumulative volume. The dissolution process for metal ions shown in Figure 5-21, which clearly reflects the PAH elution process of the soil columns by citric acid. Figure 5-19 gives the distributions of residual PAHs in the soil columns after elution by LMWOAs. Figures 5-19 and 5-21 show that during elution by LMWOAs, metal cations and PAHs in soils were released into eluates simultaneously, and the general trend that the PAH concentrations in the soil layers followed had the order bottom layer > middle layer > top layer. This indicates that the LMWOAs could activate PAHs and promote their mobilization along with the flow of organic acid solution downward through the column. Results above indicate the close correlations between the desorbed PAHs and released metal cations, and the release of both PAHs and metal cations from soils was resulted from the elution by LMWOAs.

Table 5-9 The average concentrations of metal ions (Fe, Al, Ca, Mg, and Mn) in eluates from the yellow-brown soil column at different concentrations of LMWOAs

Organic acid		Average concentration of metal ions in eluate/(mg/L)				
Type	Concentration /(mmol/L)	Fe	Al	Ca	Mg	Mn
Control	0	0.32	0.36	26.66	6.02	1.10
Citric acid	5	114.42	23.34	87.69	20.82	4.04
	10	264.03	64.56	184.33	44.71	8.88
	40	1313.78	302.95	739.65	152.59	49.94
	80	2282.74	599.98	1313.85	268.22	96.58
Malic acid	5	3.62	0.26	42.04	18.41	3.80
	10	33.47	1.97	111.12	27.85	6.11
	40	1022.62	74.43	546.19	126.56	37.78
	80	2122.82	224.19	1069.60	246.36	95.25

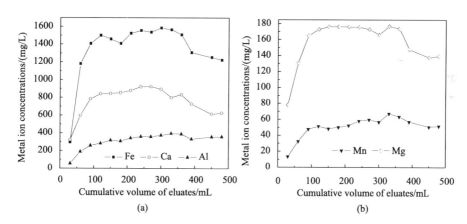

Figure 5-21 The cumulative, volume-dependent patterns of concentration curves of (a) Fe, Al, Ca, and (b) Mg and Mn ions in eluates from the yellow-brown soil column via citric acid elution at a concentration of 40 mmol/L

We analyzed the correlation between the average concentrations of metal ions in eluates and the organic acid concentrations, and the regression coefficients (R^2) are shown in Table 5-10. All the R^2 values between the average concentrations of metal ions (Fe, Al, Ca, Mg, and Mn) and citric acid concentrations were larger than 0.9824 and statistically significant. For malic acid, the R^2 values were also larger than 0.9588, again indicating significant correlations. Apparently, the dissolution of metal ions from the soil column was a direct result of the elution by LMWOAs, which could destroy organic metal ion-inorganic complexes and accordingly promote the dissolving of metal ions from soils. Furthermore, at the same concentration of LMWOAs, the promotional efficacy of the dissolution of metal ions was of the following order: citric acid > malic acid (Table 5-9). In addition, the type of LMWOA used affected the average concentrations of different metal ions in the eluates. For citric acid elution, at each citric acid concentration, the average concentrations of metal ions in the eluates were in the following order: Fe > Ca > Al > Mg > Mn. For malic acid elution, at a low concentration (5~10 mmol/L), the Ca ion had the largest concentration, but when the malic acid concentration was more than 40 mmol/L, the average concentrations of metal ions in the eluates were in the following order: Fe > Ca > Mg > Al > Mn. The results indicated that the dissolution of metal ions in soils was dependent on the type of LMWOA.

Table 5-10 The regression coefficients (R^2) between the average concentrations of metal ions (Fe, Al, Ca, Mg, and Mn) in eluates or mobilized amounts of PAHs from the yellow-brown soil column and organic acid concentrations

Parameter	Average concentrations of metal ions in eluates				
	Fe	Al	Ca	Mg	Mn
Citric acid concentration	0.9824	0.9964	0.9902	0.9954	0.9990
Malic acid concentration	0.9858	0.9820	0.9976	0.9974	0.9588
Parameter	Mobilized amounts of PAHs		Decrement in the residual concentrations of nonextractable PAHs by n-butanol	
	Phenanthrene	Pyrene	Phenanthrene	Pyrene
Citric acid concentration	0.7720	0.9200	0.8363	0.8468
Malic acid concentration	0.7124	0.8987	0.6324	0.8785

Citric and malic acids have different pKa and numbers of functional groups. The pKas of citric and malic acids are 3.13 and 3.46, respectively (Qin et al., 2004). The acidity of citric acid is greater than that of malic acid; i.e., the pH value in citric acid solution is comparatively lower. Thus, citric acid more readily promoted the dissolution of metal ions in soils compared to malic acid. Citric acid is a ternary organic acid and has three carboxyl groups and one hydroxy group, while malic acid is a binary organic acid and only has two carboxyl groups and one hydroxy group. These LMWOAs with carboxyl and hydroxy groups could compete with SOM for the binding of inorganic mineral in soils, disrupt the organic metal ion-mineral linkages (An et al., 2010), and produce dissolvable LMWOA-metal ion-mineral complexes. Elgh-Dalgren et al. (2009) found that organic acids could promote the release of SOM and that PAHs might be released into the aqueous phase along with the desorbed organic matter. However, significant correlations between the mobilized amounts of PAHs from the soil columns and organic acid concentrations were not observed (Table 5-10). The R^2 between the mobilized amounts of phenanthrene from the soil columns and malic acid concentrations was only 0.7124, which is much less than the R^2 (0.9976) between the average concentrations of Ca and the malic acid concentrations. In addition, the correlations between the decrease in the residual concentrations of nonextractable PAHs by n-butanol and organic acid concentrations were lower (Table 5). The R^2 between the decrease in the residual concentrations of nonextractable phenanthrene and malic acid concentrations was only 0.6324, indicating that the elution mechanism of PAHs in soils by LMWOAs was not in accord with that of the dissolution of metal ions.

According to the analysis above, citric acid could provide more anions for the complexing process than malic acid, which could lead to a greater efficacy of citric acid in dissolving metal ions and organic matter from soils carrying more adsorbed PAH molecules with them. However, the results presented in this work were to the contrary in that malic acid was observed to enhance a greater release of PAHs from soils than citric acid, except for at the organic acid concentration of 80 mmol/L (Table 5-8). Such differences can possibly be ascribed to the differences of soil and soil column properties. One notes that besides the mechanisms mentioned above, there might exist other mechanisms involved in the mobilization of PAHs from soils by LMWOAs. The PAH mobilization from soils into eluates could be more complicated than those of dissolution of metal ions. In all, both metal cations and PAHs in soils were simultaneously released into eluates by LMWOAs, and the release of metal cations correlated well to the release of PAHs. Unfortunately, the determinate relationship between the desorbed PAHs and released metal cations was not found in this study, and further research is necessary to elucidate the exact mechanisms.

5.5 LMWOAs enhance the release of bound PAH residues in soil

Bound residues have a direct effect on the long-term partitioning behavior, bioavailability, and toxicity of contaminants in soil (Suflita and Bollag, 1981; Pignatello, 1998; Gao et al., 2013). The formation of a bound residue is generally considered to act as a soil detoxification process by permanently binding compounds into soil matrix, and the bioavailability of bound residues is the final endpoint for the risk assessment and regulatory management of organic chemicals in the soil environment (Richnow et al., 1995; Northcott and Jones, 2000). However, recent studies found that although the available fractions of phenanthrene and pyrene in soils were removed, the clear uptake, accumulation and translocation of phenanthrene and pyrene by ryegrass (*Lolium multiflorum* Lam.) indicated a significant phytoavailability of bound-PAH residues in soils (Gao et al., 2013). However, the mechanisms involved were still under elucidation.

Recent studies have shown that LMWOAs may alter the availability of PAHs in the soil environment. Using batch equilibrium assays, Gao et al. (2010) observed that phenanthrene desorption from soil was clearly enhanced by LMWOAs. Later, Sun et al. (2012) found that the *n*-butanol extractable amounts of phenanthrene in soil increased with increasing the concentrations of root exudates; and malic acid, over alanine, serine and fructose, displayed the most significant enhancement in PAH extractability in soil. However, as to the bound PAH residues, little work is available to investigate the effects of LMWOAs as well as other root exudates on the release and availability of these bound fractions in soil. Understanding this issue will be helpful to reveal the mechanisms of phytoavailability of bound PAH residues in the soil environment.

Here, we seek to investigate the impacts of LMWOAs on the release of bound PAH residues (reference to PAH parent compounds) in soils (Gao et al., 2015b). Three LMWOAs including citric acid, oxalic acid and malic acid were experimented, since they are frequently observed with high concentrations in PAH-contaminated rhizosphere (Gao et al., 2011), and their impacts on the sorption, desorption and extractability of PAHs in soils were documented in last years (Gao et al., 2010; Sun et al., 2012). Some properties of organic acids were listed in Table 5-11. An *n*-butanol extraction procedure was used to evaluate the availability of soil PAHs (Liste and Alexander, 2002). The results of this work may provide important information on the fate of bound PAH residues in the soil environment, and will be useful in the risk assessment of bound PAH residues in soils.

Table 5-11 Some properties of the organic acids used in this study

Organic acid	Molecular formula	Molecular weight/(g/mol)	pKa
Citric acid	$HO_2CCH_2C(OH)(CO_2H)CH_2CO_2H$	192.43	3.08, 4.76, 6.40
Oxalic acid	HO_2CCO_2H	90.04	1.23, 4.19
Malic acid	$HO_2CCH(OH)CH_2CO_2H$	134.09	3.40, 5.11

Contaminated soils were collected from the A horizon (0~20 cm) around a petrochemical plant in Nanjing, China. The soil type is a TypicPaleudalf with a pH of 5.87, 13.6 g/kg soil organic carbon content, 26.3 % clay, 13.0 % sand, and 60.7 % silt. Soil samples were air dried and sieved through a 2 mm mesh. The soils containing only bound PAH residues were obtained as follows (Sabate et al., 2006; Gao et al., 2013). Solutions of dichloromethane:acetone (1 : 1, vol/vol) were added. Extraction was conducted in an ultrasonic bath. The solvent was decanted, and the samples were re-extracted with fresh solvent. This process was repeated five times. Soils were air-dried and passed through a 0.9 mm sieve, and then test soils containing only bound PAH residues were obtained. Soils containing only bound PAH residues were filled in amber glass microcosms (Macleod and Semple, 2003; Ling et al., 2010). Solution with LMWOAs and 0.5% NaN_3 was added. The control treatments without LMWOA addition received an equivalent amount of solution with 0.5% NaN_3 except for the absence of LMWOAs. Soil water content was adjusted to 30% of the soil water holding capacity. Amber glass microcosms were sealed and incubated at 25℃ for 0~80 days. Soil was then sampled, freeze-dried and passed through a 60-mesh sieve. An n-butanol extraction technique for PAHs in soil was adapted from Liste and Alexander (2002).

5.5.1 The release of bound PAH residues in soils as a function of incubation time

A method involving non-exhaustive extraction with n-butanol was selected to assess the release of bound PAH residues and PAH availability in soils (Liste and Alexander, 2002; Ling al., 2009). Eight representative PAHs, including naphthalene (NAP), acenaphthene (ACP), fluorene (FLU), phenanthrene (PHE), fluoranthrene (FLA), pyrene (PYR), benzo(a)anthrancene (BAA) and benzo(k)fluoranthrene (BKF), with different physiochemical properties were investigated in this study. The concentrations of *n*-butanol-extractable PAHs (BEPAHs) in soil prepared with bound PAH residues as a function of soil incubation time (0~80 days) are shown in Figure 5-22. No BEPAHs were observed at day 0. However, the concentrations of BEPAHs in soils increased with prolonged soil incubation time, indicating a significant release of bound PAH residues in soils. The concentration of total BEPAHs after 80 days was 14.3 mg/kg on a dry weight basis. The concentrations of individual BEPAH varied extensively among test PAHs but all were less than 4.2 mg/kg. In general, the concentrations of BEPAHs in soils after 10~80 days incubation followed the descending order of FLA > PYR > FLU > ACP, BKF > PHE > BAA > NAP. The above results suggest that bound PAH residues in soils could be released and become partially extractable after incubation, and a longer incubation time would result in a greater release of bound PAH residues and a larger portion of extractable PAHs remaining in the soil environment.

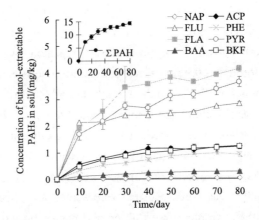

Figure 5-22 Concentrations of *n*-butanol-extractable PAHs in soil prepared with bound PAH residues as a function of incubation time. Error bars represent standard deviations

The clear increase of BEPAH concentrations in soils prepared with bound PAH residues over the extended incubation time (0~80 days) indicated a significant release of bound PAH residues from soils. Some studies have also suggested that bound residues should not be considered to be permanently bound as the potential for partial reversibility always exists, and pollutants may subsequently be released from soil by the continuous turnover of soil organic carbon or undergo further entrapment or binding as a result of humification and diagenesis (Northcott and Jones, 2000). For example, Khan and Ivarson (1981) found that soil-bound ^{14}C-labeled residues were released by microbes from an organic soil treated with ^{14}C-ring-labeled prometryn, and 27% of the total ^{14}C in soil was extractable after a 22 days incubation. However, most literature has focused on the release of bound herbicide and pesticide residues in the soil environment, and to our knowledge, this is the first investigation that has directly observed the release of bound PAH residues in soils. Since microbial activity was inhibited during incubation, the release of bound PAH residues from soils may dominantly be ascribed to the chemical processes. As reported, the solvent-nonextractable PAHs in soils were associated with the humic/fulvic acid and humin fractions, which serve as the dominant sink for bound PAH residues (Nieman et al., 1999). With the exceedingly slow diffusion of a PAH through these solid phase of SOM and to its outside would it then become available. In addition, there is possibility that these solid fractions of SOM may partially release and become soluble fractions in test soils after incubation for days (Alexander, 2000), resulting in the observed release of soil bound PAH residues.

5.5.2 LMWOA-enhanced release of bound PAH residues in soil

The amounts of total BEPAHs in test soils prepared with bound PAH residues as a function of LMWOA concentrations (0~100 mmol/kg) after 40 days and 80 days are given in Figure 5-23. The addition of LMWOAs significantly increased the concentrations of total BEPAHs in a LMWOA-concentration dependent manner, and higher concentrations of LMWOAs added resulted in larger amounts of BEPAHs in soils. For example, the concentrations of total BEPAHs in soils increased from 11.8 mg/kg and 14.3 mg/kg to 66.9 mg/kg and 73.5 mg/kg with an increase in the citric acid concentration from 0 mmol/kg to 100 mmol/kg after 40 days and 80 days, respectively. Similar trends were also observed for the treatments using oxalic or malic acid. These findings indicate that the LMWOAs tested significantly enhanced the release of bound PAH residues and enlarged the extractable amounts of PAHs in aging soils. As with the control treatment with no addition of LMWOAs (Figure 5-22), a longer time period (80 versus 40 days) increased the release of bound PAH residues in soils when LMWOAs were added. However, the total BEPAHs differed clearly among the tested LMWOAs. The amounts of total BEPAHs in soils with LMWOAs added at test concentrations (0~100 mmol/kg) always followed the descending order of citric acid > oxalic acid > malic acid, regardless of the incubation time (Figure 5-23).

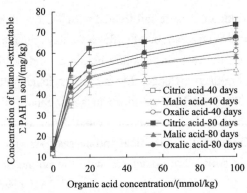

Figure 5-23 Concentrations of *n*-butanol-extractable ΣPAH in soils prepared with bound PAH residues after incubation for 40 days and 80 days as a function of the concentrations of organic acids added to the soils. Error bars represent standard deviations

The enhancement ratio (r, %) of BEPAHs in soils following the addition of LMWOAs was calculated as follows

$$r = (C_i - C_o) / C_o \times 100\%$$

where, C_i and C_o were the BEPAH concentrations in soils with and without the addition of LMWOA, respectively. Higher r values indicate a greater potential for the organic acids to enhance PAH release and increase extractability from soils. The calculated r values for total BEPAHs in soils following the addition of LMWOAs are shown in Figure 5-24. The r values for BEPAHs by citric acid, oxalic acid and malic

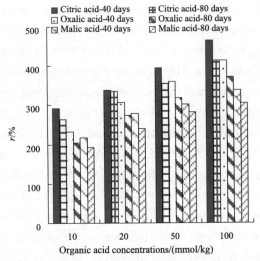

Figure 5-24 Enhancement ratio (r, %) of *n*-butanol-extractable ΣPAH in soils prepared with bound PAH residues after 40 days and 80 days following the addition of organic acids at different concentrations. Error bars represent standard deviations

acid at 10~100 mmol/kg were 293%~465%, 232%~413% and 217%~339% after 40 days and 264%~413%, 204%~373% and 192%~306% after 80 days, respectively. r values significantly increased with increasing organic acid concentrations, and followed the descending order of citric acid > oxalic acid > malic acid. In addition, the r values of BEPAHs in soils with LMWOAs after the 40-day incubation were always higher than after 80 days, indicating that a longer incubation time increased the amount extracted (Figure 5-23), but reduced the enhancement ratio of the release of bound PAH residues in soils following the addition of LMWOAs.

The BEPAH concentrations of each individual PAH in soils prepared with bound PAH residues after 40 days and 80 days incubation as a function of LMWOA concentration are given in Figures 5-25. As with the trend of ΣPAH, the concentrations of individual BEPAH always increased with increasing concentration of LMWOAs (0~100 mmol/kg), irrespective of incubation time, and PAH and organic acid species. Among the three LMWOAs tested, the BEPAH concentrations generally descended in the order of citric acid > oxalic acid > malic acid. For the different PAHs, the BEPAH concentrations in soils with the same LMWOA concentrations applied decreased in the order of PYR > FLA > ACP > BKF ≈ FLU > PHE > BAA > NAP. In addition, the BEPAH concentrations in soils with LMWOAs added after the 80-day incubation were always higher than after 40 days, indicating that a longer incubation time led to greater PAH release from soils. For example, the n-butanol-extractable concentrations of PYR, FLA, ACP, BKF, FLU, PHE, BAA and NAP in soils with 100 mmol/kg oxalic acid were 20.5 mg/kg, 16.5 mg/kg, 7.46 mg/kg, 5.81 mg/kg, 5.38 mg/kg, 3.97 mg/kg, 1.03 mg/kg and 0.16 mg/kg after 40 days, and 24.6 mg/kg, 17.0 mg/kg, 8.56 mg/kg, 5.88 mg/kg, 6.09 mg/kg, 4.28 mg/kg, 1.19 mg/kg and 0.16 mg/kg after 80 days, respectively.

The r (%) values of the individual BEPAH concentrations in soils following the addition of LMWOAs are shown in Figure 5-26. Enhancement ratio (r) values were always positive when organic acids were added to soils, and increased with increasing organic acid concentration (0~100 mmol/kg). The r values of the butanol-extractable concentrations of PYR, FLA, ACP, BKF, FLU, PHE, BAA and NAP in soils with 10~100 mmol/kg organic acids were 391%~745%, 164%~394%, 275%~635%, 197%~565%, 54%~139%, 236%~477%, 139%~406% and 356%~531% after 40 days, and 299%~605%, 131%~340%, 295%~700%, 172%~469%, 58%~126%, 146%~400%, 139%~407% and 320%~496% after 80 days, respectively. The observed r values for the same treatments after the 80 days incubation were always smaller than those after 40 days, although more PAHs were released from the soils after 80 days(Figure 5-25). In addition,

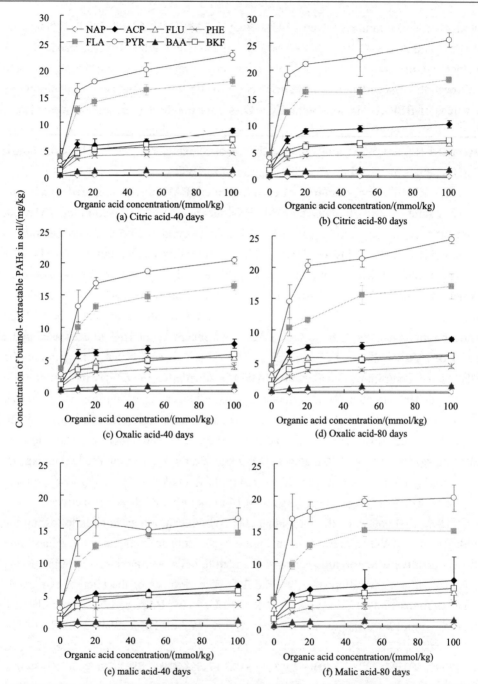

Figure 5-25 Concentrations of *n*-butanol-extractable PAHs in soils prepared with bound PAH residues after 40-day and 80-day incubation as a function of the concentration of organic acids added. Error bars represent standard deviations

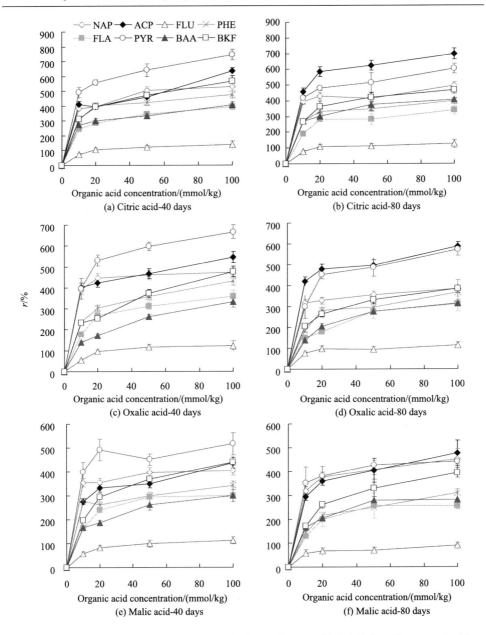

Figure 5-26　The enhancement ratio (r, %) of n-butanol-extractable PAHs in soils prepared with bound PAH residues after 40 days and 80 days following the addition of organic acids at different concentrations. Error bars represent standard deviations

citric acid generated the largest enhancement of (with the biggest r values) the PAH release from soil bound PAH residues, while malic acid generated the lowest release of the three organic acids tested.

The enhanced release of bound PAH residues in soils by the LMWOAs was a novel finding of this study, which can be attributed to LMWOA-influenced microbial interactions and chemical reactions. As described previously, a partitioning into SOM and then a slow diffusion within the organic matrix or entrapment in micropores within soil aggregates are the primary mechanisms for the formation of bound HOC residues (Pignatello, 1998; Nieman et al., 1999; Chilom et al., 2004; Gao et al., 2013). The SOM serves as the main pool of bound residues of HOCs, such as PAHs, in soils (Chiou et al., 1998). However, SOM itself is subject to alteration by chemical and biotic processes (Käcker et al., 2002). It has been well documented that metallic cations in soils form complexes with functional groups of soil organic molecules, and this association leads to the formation of "bridges" between minerals and the SOM (Saison et al., 2004). LMWOAs can bind metal cations. When released into the rhizosphere, LMWOAs can dissolve the metal cations and break the bridges between the solid surface of soil particles and the SOM, resulting in the release of SOM into solution (Gao et al., 2010). This is supported by our previous findings that LMWOAs significantly increase the concentrations of metal cations in soil solution (Sun et al., 2012). The dissolution of some soil minerals by root-secreted organic acids has been reported (Drever and Stillings, 1997), which theoretically also breaks the mineral-SOM association and reduces SOM. As a consequence, the release of SOM into soil solution results in the associated bound PAH residues in soils becoming extractable and bioavailable. In addition, enhanced microbial activity in the rhizosphere due to root exudates, including LMWOAs, has been reported (Grayston et al., 1997; Bais et al., 2006), which in theory can promote the release of bound PAH residues in soils by microbial digestive extraction and excretion, SOM metabolism and bioturbation (Northcott and Jones, 2000). However, because soil microbial activity was inhibited in this investigation, the observed LMWOA-enhanced release of bound PAH residues in soils was predominantly due to the chemical reactions involving LMWOAs, as stated above. It should be noted that citric acid, which has three carboxyl groups, always resulted in a greater release of bound PAH residues in soils than oxalic and malic acids, which have two carboxyl groups; this might be attributed to citric acid having greater potential for chelation with metal cations and promoting the dissolution of soil minerals (Gao et al., 2003, 2010).

PART II PLANT

Chapter 6   Uptake, accumulation and translocation of PAHs in plants

Soil and water contamination are long-term environmental problems generated by anthropogenic activities of the last decades. Natural and xenobiotic organic pollutants present in soil, water and air may be taken up by plants, which are the major pathways for toxic substances into the food chain (Paterson and Mackay, 1994; Trapp and Matthies, 1995; Voutsa and Samara, 1998; Gao et al., 2013b). Since plants form the basis of human and animal food webs, potentially harmful organic contaminants could find their way into human and animal populations via this route. Clearly, increased understanding of how plants take up and accumulate organic contaminants from the environment will have considerable benefits for risk assessment purposes (Binet et al., 2000; Fryer and Collins, 2003).

During the last decades, there has been a considerable interest in the uptake of organic chemicals including PAHs by plants (Topp et al., 1986; Wang and Jones, 1994; Mattina et al., 2003). Plant can be exposed to contaminants in different ways. Active or passive uptake by the roots occurs, possibly followed by transport along with the transpiration stream. Shoot uptake from air occurs by wet and/or dry deposition on the above-ground parts and possibly absorption and translocation over the cuticles. Another possible way as speculated by researchers is the uptake and transport in oil cells that might be found in oil-containing plants such as carrots and cress (Topp et al., 1986; Polder et al., 1995). Plant uptake and distribution have been shown to be dependent on the physical-chemical properties of the organic chemicals, the characteristics of the soil and water, and the plant species and physiology including properties such as lipid or water contents and transpiration rates (Paterson and Mackay, 1994). Plant uptake is also one approach involved in rhizoremediation for organic contaminated sites, and information about plant PAH concentrations is essential to predict the effectiveness of a rhizoremediation operation (Sung et al., 2002; Gao and Zhu, 2003; Joner and Leyval, 2003; Gao and Zhu, 2005).

In this chapter, the uptake pathway of PAHs in plants was clarified. The accumulation and translocation of PAHs in plants with different compositions were investigated. The PAH uptake by plants from soil and water was comparably evaluated.

Results are useful in evaluating human exposure risks of PAH-contaminated crops and in developing appropriate strategies for the rhizoremediation of PAH-contaminated sites.

6.1 Uptake pathways of PAHs in plants

Understanding the predominant pathways for plant uptake will enable us to protect human and ecological health. Dry deposition from the atmosphere (Simonich and Hites, 1994; Howsam et al., 2001), and transfer from the soil to root and shoot (Ryan et al., 1988; Gao and Zhu, 2004) have both been identified as significant pathways of crop contamination with organic pollutants. Few studies exist where all potential contamination pathways have been isolated. Many papers simply compare a bioconcentration factor (BCF) for the plant material based on a soil concentration. It is only by pathway identification we can appreciate the most important variables to measure when gathering data to calibrate and validate plant uptake models.

In the past decades, several researchers have documented the uptake pathways of organic chemicals by plants. Welsch-Pausch et al. (1995) investigated the main pathways of ryegrass contamination by dioxins and furans and reported that dry deposition was the most significant, little root uptake and subsequent translocation of these compounds would be expected because of their high K_{ow} (Collins et al., 2006). Volatilization and subsequent dry deposition was a significant pathway for chlorobenzenes and was related to their K_{oa}, as has been reported for PCBs (Wang and Jones, 1994; Pier et al., 2002), while root uptake and transport in the transpiration stream was found to be the dominant pathway for the plant accumulation of pyrene and phenanthrene from soil which was correlated to the K_{ow} of the pollutant (Gao and Zhu, 2004). The latter pathway has been investigated by a number of workers and the transport from the root to the shoot is governed by a bell shaped relationship with a maximum lg $K_{ow} \approx 2$ (Briggs et al., 1983; Burkenand Schnoor, 1998). Plant composition is frequently a parameter within plant uptake models, with retention of organic pollutants being calculated from the lipid content of the plant and the K_{ow} of the chemical under investigation (Trapp and Matthies, 1995; Chiou et al., 2001). When the maximum potential uptake by plant material was calculated using a composition partition model good agreement was reported for lindane and chlorinated solvents (Li et al., 2005).

PAHs are frequently detected, toxic, industrial pollutants that may potentially accumulate in plant material and subsequently contaminate food chains. They have a range of physicochemical properties which will result in potentially different contamination pathways for individual PAH accumulation by the plant. For example, Cousins and Mackay (2001) suggested that uptake from the atmosphere is the important pathway for chemicals with lg K_{oa}>6 and lg K_{aw}>−6 (K_{aw}=dimensionless air-water partition coefficient), and uptake from the soil with subsequent translocation to the leaves is important for chemicals with lg K_{ow}<2.5 and lg K_{aw}<−1. Here, we seek to investigate the primary pathways of accumulation of PAHs in white clover (*Trifolium repens*) (Gao and Collins, 2009).

Six PAHs were experimented and their physiochemical properties were listed in Table 6-1. Plants were first germinated and grown in acid washed sand, and then were transferred to glass jars containing 1/4 strength Hoagland's solution. The treatments imposed were designed to quantify the pathways of uptake of PAH. Treatment A—passive root uptake, Treatment B—active root uptake and translocation, Treatment C—influence of transpiration on root uptake and translocation, Treatment D—phloem mobility, Treatment E—volatilization and shoot deposition. Each solution was spiked with a mixture of the six PAHs (Figure 6-1). Treatments A, B and C were also used to investigate the impact of growth solution concentration and time respectively on the uptake dynamics of PAHs in white clover. After 6 days the plants were harvested and washed with de-ionized water, dried with tissue paper, freeze dried, and PAH-detected. Data are expressed as a root concentration factor (RCF) where RCF = root concentration (mg/g)/solution concentration (mg/mL), shoot concentration factor (SCF) where SCF = shoot concentration (mg/g)/solution concentration (mg/mL) and as a transfer factor (TF) where TF = shoot concentration (mg/g)/ root concentration (mg/g).

Table 6-1 Selected physicochemical properties and analytical recoveries of PAHs used in this study

PAHs	Abbreviation	lgK_{ow}	S_w/(mg/L)	lgK_{aw}	lgK_{oa}	Extraction efficiency	
						root	shoot
Naphthalene	Nap	3.37	31.0	−1.75	5.13	100.6	67.3
Acenaphthene	Ace	3.92	3.8	−2.31	6.39	99.1	73.1
Fluorene	Flu	4.18	1.9	−2.50	6.68	97.3	74.0
Phenanthrene	Phe	4.57	1.1	−2.88	7.45	94.5	76.3
Fluoranthene	Fluan	5.22	0.26	−3.38	8.60	84.4	80.1
Pyrene	Pyr	5.18	0.13	−3.43	8.61	84.5	85.9

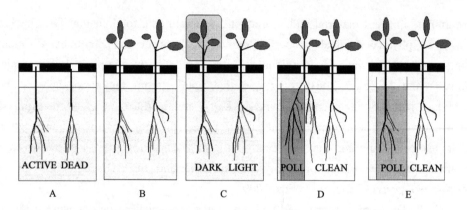

Figure 6-1 Experimental treatments to determine primary pathways of uptake of PAH into plants. Tint light coloring indicates PAH spiked solution. Black coloring indicates PAH-free solution. A: Bare roots in spiked solution; A_{ACTIVE}, shoot removed at start of the experimental period; A_{DEAD}, root killed by acid treatment. B: Plants grown in spiked solution with different PAH concentrations. C: Plants grown in spiked solution, plant shoots in the light C_{LIGHT} and half darkness C_{DARK}. D: Half plants grown in clean solution D_{CLEAN}; others have half of their roots in spiked solution and half in clean solution D_{POLL}; all the plants sharing the same air conditions. E: Half plants grown in spiked E_{POLL} and half in clean E_{CLEAN} solution; all the plants sharing the same air conditions

6.1.1 Root uptake of PAHs

The accumulation of PAHs in roots increased with increasing K_{ow} of the compound. The model of Ryan et al. (1988) which describes the root uptake of organic chemicals from solution

$$RCF = 10^{0.77 \lg K_{ow} - 1.52} + 0.82$$

has been used widely in plant uptake models for risk assessments (Pier et al., 2002), underestimated the root accumulation observed here. When the same function

$$RCF = 10^{a \lg K_{ow} - b} + c$$

was used to fit the data the constants b and c were not significant, with a=0.74, this accounted for 65% of the variation in the data. However use of this model would result in a negative RCF for those compounds with $\lg K_{ow}$<3.8, so c was set to 5.5 the mean value for the A_{ACTIVE} root systems with $\lg K_{ow}$<3.8 in these experiments. This alteration did not affect the value of a, or the proportion of variance the fitted function accounted for.

Within the treatments there were significant differences in the RCF between B

and A_{DEAD} for all PAHs ($p<0.01$), there was a higher root accumulation of napthalene in B c.f. A_{ACTIVE} ($p<0.05$), but not for other PAHs ($p>0.05$) (Figure 6-2). There were no differences in RCFs between those plants with shoots grown in the light compared to those with shoots in the dark (Treatment C). These findings indicate that an active root is required to accumulate PAHs, it is not solely a sorption process as indicated by many workers, and that a transpiration stream flux enhances root uptake for more water-soluble compounds such as napthalene. There was no phloem mobility of PAHs (Treatment D) as none were recovered from the half of root growing in the clean solution. This would support the paradigm of phloem mobile compounds needing a lg $K_{ow}<0$ or an acid dissociation constant >-6 and lg $K_{ow}<3$ (Bromilow and Chamberlain, 1995).

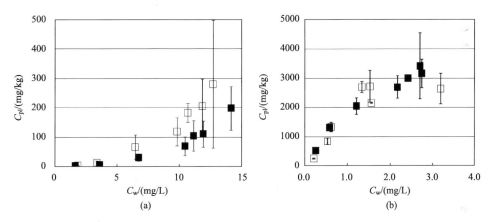

Figure 6-2 Uptake concentrations of PAHs in plant roots with varying solution concentration (a) naphthalene (b) phenanthrene. C_{pl} = plant concentration, C_w = solution concentration. Closed symbols A_{ACTIVE}, open symbols B. Error bars are standard deviations

6.1.2 Shoot accumulation of PAHs

Shoot accumulation of PAH also increased with increasing K_{ow}, Nap<Ace< Flu=Phe<Fluan ($p<0.05$). The TF showed the reverse trend Nap>Ace>Flu>Phe> Fluan ($p<0.05$) demonstrating the importance of water solubility in the transfer of PAH from the root to the shoot. Others have expressed this transfer as the transpiration stream concentration factor which is also seen to decline with increasing lg K_{ow} beyond a value of 2 (Briggs et al., 1983; Hsu et al., 1990). As observed for the RCF there was no difference in SCF between C_{LIGHT} and C_{DARK} indicating that over the relatively short duration of this experiment shading did not significantly affect the accumulation of PAH in the shoot, or more likely it was ineffective in reducing the transpiration flux. Future experiments may require a longer period to reveal differences as a consequence of increased transpiration, as over the 6 days experimental period equilibrium had not been obtained for any of the PAHs as demonstrated by the increase in the TF with time (Figure 6-3). In similar studies with ryegrass, shoot concentrations of hexachlorobenzene (lg K_{ow}=5.5), tetrachloroethylene (lg K_{ow}=3.38), and trichloroethylene (lg K_{ow}=2.53) attained equilibrium after >200 h, 80 h and 60 h, respectively (Li et al., 2005). The longer equilibrium times for higher K_{ow} compounds have been attributed to the kinetic limits on the uptake of these compounds, related to the their low water solubility and hence depleted concentration in the transpiration stream relative to the root (Trapp and Matthies, 1995).

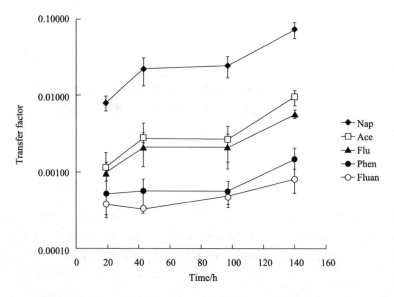

Figure 6-3 Alteration in the transfer factor from root to shoot of PAH with time. Pyrene data are not reported for the shoots as an unknown compound co-eluted with this PAH preventing accurate quantification. Error bars are standard deviations

For both D_{POLL} and E_{POLL} there was significant accumulation in the shoot compared to clean treatments ($p<0.001$) (Figure 6-4), but D_{CLEAN} and E_{CLEAN} treatments had higher shoot concentrations than plants grown in a clean environment. When the two pathways of contamination were separated root to shoot transfer via transpiration was dominant for Nap, Ace, Flu, and Phen and dry deposition from the surrounding atmosphere dominated for Fluan. The potential significance of dry deposition following volatilization has been previously reported (Smith et al., 2001). There was no significant correlation with chemical properties in our study, but chemicals with lg K_{oa} > 9 and lg K_{aw} <−3 have been proposed to have a significant proportion of their contamination via the soil air plant pathway, while others have stated that chemicals with lg K_{aw} < −4 may are predominantly deposited via this route (Ryan et al., 1988). Fluan (lg K_{oa}> 8.6 and lg K_{aw} <−3.38) is the closest to these physicochemical properties. The high contribution of direct aerial deposition from volatilization to plant contamination is noteworthy and supports the proposal that this is an important pathway for the accumulation of organic chemicals by vegetation from polluted soils (Harrad et al., 2006). This pathway is often ignored in studies used to parametrize models for the uptake of organic chemicals by crops.

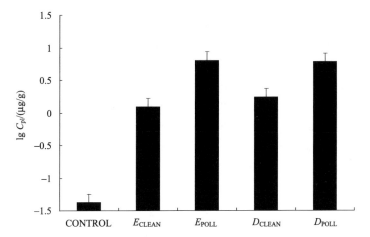

Figure 6-4 Difference in shoot concentration (C_{pl}) for mean sum of PAH concentration between clean and polluted treatments. D_{POLL} plants grown with half roots in contaminated solution and shoot in contaminated air; D_{CLEAN} plants grown with roots in clean solution and shoot contaminated air; E_{POLL} plants grown with roots in contaminated solution and shoots contaminated air; E_{CLEAN} plants grown with roots in clean solution but shoots in contaminated air, CONTROL plants grown in clean solution and air. Error bars are standard deviations

The accumulation in the shoot was driven by the transpiration stream flux for all PAHs (r^2 > 0.65), the gradient calculated from the fitted linear functions for each individual PAH correlated with the water solubility of the individual chemicals (r^2=0.998, p < 0.001), but not their lg K_{ow} (r^2=0.48, p > 0.1). This finding arose because the plant material had not attained equilibrium with the contaminants over the short time scale of the study (Figure 6-3). Where equilibrium or near equilibrium conditions exist the relation with lg K_{ow} is likely to be improved.

6.1.3 Uptake time and PAH concentration influence their uptake by plants

The RCF declined with time for Nap, did not alter for Ace and Flu, and increased throughout the experimental period for Phe, Fluan, and Pyr but increased at a slower rate from 2 days. The increasing time to equilibrium with increasing K_{ow} and declining accumulation rate has been observed previously (Li et al., 2005). The SCF increased more rapidly with time for those PAH with lower K_{ow}. These findings would be supported by transpiration stream concentration factor relationships reported by others where transport to the shoot declines with lg $K_{ow}>3$ (Briggs et al., 1983; Burken and Schnoor, 1998). The increasing SCF over time results from the larger volume of water transpired with time which drives the flux of chemical into the shoot. The combination of these processes resulted in an increasing TF over the duration of the experiment, with higher TFs for those PAH with higher S_w and lower K_{ow} (Figure 6-3).

The RCF increased with increasing concentration for all PAHs (Figure 6-5). With the exception of naphthalene all the PAHs exhibited an asymptote at the higher concentrations. Partition coefficients of the individual PAH to plant material (K_{pl}), were calculated from a lipid composition model

$$C_{eq} = C_w (f_{lip} K_{ow})$$

$$K_{pl} = C_{eq} / C_{sol}$$

where C_{eq}=equilibrium concentration of PAH in the plant; C_{sol}=solution concentration of PAH; f_{lip}=fraction of plant lipid (0.055 for shoot and 0.028 for root on a dry weight basis); K_{pl} was also calculated using a Freundlich model, and a sorption model from fits to the linear portion of the data (Figure 6-5 and Table 6-2). The three models exhibited some deviation, with the lipid model predicting higher accumulation of naphthalene, but lower accumulation for the remaining PAH compared to the sorption and Freundlich models. The Freundlich model predicted very high K_{pl} for pyrene. The potential reasons for this are the lower K_{ow} compounds are more likely to be subject to metabolism within the plant; second deviations between octanol and plant lipids, other workers have found triolein to be a better surrogate; and third the lipid extraction was not exhaustive, a multiphase extraction has been proposed to extract all lipid components from plant tissue (Zhu et al., 2007).

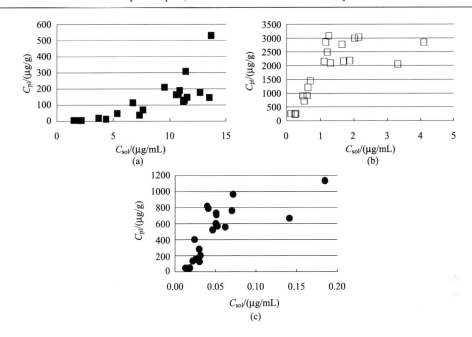

Figure 6-5 Influence of solution concentration (C_{sol}) on root concentration (C_{pl}) of (a) napthalene, (b) phenanthrene, and (c) pyrene

Table 6-2 Calculated partition (K_{pl}) of PAH from solution to plant components using a lipid composition model and Freundlich fit to experimental data and a sorption model from the linear portion of the data

		K_{pl}					
		Nap	Ace	Flu	Phen	Fluan	Pyr
Root	Lipid composition model	65.6	232	423	1040	4650	4240
	Sorption model	19.4	277	650	2290	8230	15 700
	Freundlich model	1	111	552	1700	12 800	230 000
Shoot	Lipid composition model	165	582	1059	2600	11 617	10 595
	Sorption model	165	582	1059	2600	11 617	10 595
	Freundlich model	0.09	1.02	4.04	7.18	51.3	

Increasing the concentration of the PAH reduced the transfer from the root to the shoot for all PAHs. This was a consequence of reduced transpiration, which declined as the PAH concentration of the spiked solution increased ($r^2=0.58$, $p<0.001$), the Freundlich model could not therefore be used to calculate the K_{pl}. Using the SCF from

across all treatments, which was equivalent to the lowest concentration used in treatment B, to derive a K_{pl} a large divergence can be seen with the lipid model. The main reason for this is that the shoots had not attained equilibrium as discussed previously. In addition the toxic action of the PAH may reduce shoot uptake, whereas metabolism of the compounds after uptake will also decrease shoot concentrations.

In summary, root uptake was primarily an absorption process, but a component of the contamination was a result of the transpiration flux to the shoot for higher solubility compounds. The root contamination can be easily predicted using a simple relationship with K_{ow}, although if a composition model was used based on lipid content, a significant under prediction of the contamination was observed. Shoot uptake was driven by the transpiration stream flux which was related to the solubility of the individual PAH rather than the K_{ow}. However, the experiment was over a short duration, 6 days, and models based on K_{ow} may be better for crops grown in the field where the vegetation will approach equilibrium and transpiration cannot easily be measured. A significant fraction of the shoot contamination resulted from aerial deposition derived from volatilized PAH. This pathway was more significant for compounds approaching lg K_{oa}>9 and lg K_{aw}<−3. The shoot uptake pathways need further investigation to enable them to be modeled separately. There was no evidence of significant systemic transport of the PAH, so transfer outside the transpiration stream is likely to be limited.

6.2 Accumulation and translocation of PAHs in plants with different compositions

The aim of this section was to comparatively investigate the accumulation and translocation of PAHs in plants with different compositions (Gao and Zhu, 2004). Phenanthrene and pyrene were experimented as representative PAHs. Twelve plants, including three-colored amaranth (*Amaranthus tricolar* Linn., P1), flowering Chinese cabbage (*Brassica Parachinensis* Bailey, P2), ryegrass (*Lolium multiflorum* Lam., P3), radish (*Raphnus sativus* L., P4), water spinach (*Lpomoea aquatica* Forsk, P5), green soybean (*Glycine max* Merr., P6), kidney bean (*Phaseolus vulgaris* L., P7), pakchoi (*Brassica chinensis* L., P8), broccoli (*Brassica oleracea* L., P9), sinage (*Spinacea oleracea* L., P10), capsicum (*Capsicum annuum* L., P11), and eggplant (*Solanium melongena* L., P12), were vegetated in PAH-spiked soils in a greenhouse. The initial concentrations of phenanthrene and pyrene in soils were listed Table 6-3. Soils and plants were destructively sampled after 45 days treatment since sowing, and then detected for PAH concentrations (Simonich and Hites, 1994; Kipopoulou et al., 1999). The plant compositional effects on plant accumulation were evaluated.

Table 6-3   Initial concentrations of phenanthrene and pyrene in treated soils (mg/kg, on a dry weight basis)

Soil No.	S0*	S1	S2	S3	S4	S5	S6	S7	S8
Phenanthrene	ND	7.45 (0.23)	13.1 (0.90)	33.8 (2.48)	65.8 (4.53)	102 (5.29)	133 (7.39)	261 (13.5)	457 (22.2)
Pyrene	ND	8.01 (0.42)	17.2 (0.64)	48.7 (3.76)	92.7 (4.97)	131 (8.16)	172 (9.21)	290 (13.2)	489 (19.9)

Note: * Control uncontaminated soil. ND—not detectable. Data in bracket were standard deviations; $n = 3$.

6.2.1   Accumulation of PAHs in roots

Concentrations of phenanthrene and pyrene in roots grown in unspiked control soil were not detectable. This indicated that phenanthrene and pyrene accumulated by shoot from air were seldom transported to root, and these chemicals in roots grown in various spiked soils derived of root uptake and accumulation (Schroll et al., 1994). Off-take of phenanthrene and pyrene by roots was subsequently accumulated in roots or probably translocated to the upper part of plants, and a portion of them might be metabolized in plants (Trapp et al., 1990).

Root concentrations of phenanthrene and pyrene, on a dry weight basis, as a function of their soil concentrations were listed in Table 6-4. Root accumulations of these PAHs increased with the increment of their soil concentrations. As to the same soil-plant treatment, concentrations of pyrene in roots of plants, flowering Chinese cabbage as representative, generally were 6.6~36.2 times higher than those of phenanthrene. It was notable that even roots grown in the lightly spiked soils with initial phenanthrene of 7.45 mg/kg and pyrene of 8.01 mg/kg (S1), root contamination was obvious, and the respective root concentrations of phenanthrene and pyrene increased to 0.78 mg/kg and 5.12 mg/kg. While in heavily spiked soils such as S8, roots of flowering Chinese cabbage were grievously contaminated, and phenanthrene and pyrene concentrations in roots were 11.9 mg/kg and 428 mg/kg, respectively.

Table 6-4 Root concentrations (C_{pd}, mg/kg) and concentration factors (RCFs) of phenanthrene and pyrene for flowering Chinese cabbage as a function of their soil concentrations (C_s, mg/kg) after 45 days. Concentrations of phenanthrene and pyrene in soils, roots and shoots were on a dry weight basis

Soil No.	C_s-ph /(mg/kg)	C_s-py /(mg/kg)	Roots			
			C_{pd}-ph /(mg/kg)	C_{pd}-py /(mg/kg)	RCF-ph	RCF-py
S0	ND	ND	ND	ND	—	—
S1	0.85(0.10)	2.71(0.28)	0.78(0.09)	5.12(0.55)	0.91(0.09)	1.90(0.21)
S2	0.93(0.08)	2.81(0.34)	0.91(0.09)	7.14(0.70)	0.98(0.08)	2.54(0.28)
S3	2.81(0.34)	11.3(1.14)	2.72(0.25)	21.5(3.20)	0.97(0.09)	1.91(0.21)
S4	6.71(0.81)	34.0(5.01)	3.27(0.37)	67.0(5.09)	0.49(0.04)	1.97(0.18)
S5	8.60(0.75)	41.9(3.77)	3.85(0.42)	82.2(6.49)	0.45(0.04)	1.96(0.22)
S6	9.42(1.02)	58.8(6.08)	4.24(0.38)	112(9.96)	0.45(0.05)	1.90(0.15)
S7	13.8(1.48)	123(14.6)	5.7(0.68)	207(16.8)	0.41(0.03)	1.60(0.14)
S8	36.9(4.33)	301(39.2)	11.9(1.65)	428(36.8)	0.32(0.03)	1.42(0.16)

Note: C_s-ph and C_s-py were the soil concentrations of phenanthrene and pyrene (mg/kg); C_{pd}-ph and C_{pd}-py were plant concentrations of phenanthrene and pyrene (mg/kg); RCF-ph and RCF-py were root concentration factors of phenanthrene and pyrene. ND—not detectable. Data in bracket were standard deviations; $n=3$.

Great variations of root phenanthrene and pyrene concentrations were observed among different plant species (Table 6-5). Concentrations of phenanthrene and pyrene in various roots grown in S6 were 0.60~6.72 mg/kg and 13.1~199 mg/kg, respectively. Green soybean exhibited the highest root concentrations of phenanthrene (6.72 mg/kg) and pyrene (199 mg/kg), while capsicum contained the lowest portion of these compounds (phenanthrene of 0.60 mg/kg and pyrene of 13.1 mg/kg). The growth conditions of various plant species grown in S6 were identical. Thus the disparity of root uptake of these PAHs would come from plant properties.

Relationships between root concentrations of phenanthrene and pyrene and root compositions, including root lipid and water contents, were examined by regression analysis. R^2 values from linear regressions were always below 0.23 ($n=12$) for correlations of root water contents and PAH concentrations. Whereas significantly positive correlations ($p<0.05$) were found between root lipid contents and root phenanthrene or pyrene concentrations, and R^2 values from linear regressions were

Table 6-5 Root concentrations (C_{pd}) and concentration factors (RCFs) of phenanthrene and pyrene, on a dry weight basis, for plants grown in spiked soils with initial phenanthrene concentration of 133 mg/kg and pyrene of 172 mg/kg after 45 days

Plant No.	Phenanthrene			Pyrene		
	C_{pd}/(mg/kg)	C_s/(mg/kg)	RCFs	C_{pd}/(mg/kg)	C_s/(mg/kg)	RCFs
P1	1.95(0.12)	16.4(2.41)	0.12(0.02)	45.1(4.32)	63.3(7.23)	0.71(0.07)
P2	4.24(0.38)	9.42(1.02)	0.45(0.05)	112(9.96)	58.8(6.08)	1.90(0.15)
P3	1.85(0.21)	13.4(1.53)	0.14(0.02)	19.2(2.33)	51.1(5.04)	0.38(0.04)
P4	2.36(0.19)	8.71(0.93)	0.27(0.02)	82.9(6.89)	55.0(5.07)	1.51(0.11)
P5	0.85(0.09)	15.9(1.55)	0.05(0.01)	38.6(3.66)	65.0(6.04)	0.59(0.04)
P6	6.72(0.52)	10.0(1.03)	0.67(0.06)	199(13.4)	44.9(5.04)	4.44(0.36)
P7	2.53(0.31)	14.7(1.84)	0.17(0.01)	48.5(5.27)	61.1(5.86)	0.79(0.09)
P8	2.38(0.22)	12.1(1.04)	0.20(0.02)	74.8(6.91)	57.0(6.23)	1.31(0.12)
P9	0.81(0.07)	10.9(1.03)	0.07(0.01)	16.1(2.01)	56.5(5.35)	0.29(0.03)
P10	2.03(0.16)	15.7(2.06)	0.13(0.01)	20.5(1.87)	55.3(5.14)	0.31(0.04)
P11	0.60(0.04)	12.0(1.05)	0.05(0.01)	13.1(1.06)	57.5(6.46)	0.23(0.02)
P12	0.85(0.09)	10.9(1.03)	0.08(0.01)	14.5(1.72)	52.4(5.74)	0.28(0.03)

Note: C_s was the soil concentrations of phenanthrene and pyrene; C_{pd} was the root concentrations of phenanthrene and pyrene. RCF was root concentration factor. P1~P12 were plant No. Data in bracket were standard deviations; $n = 3$.

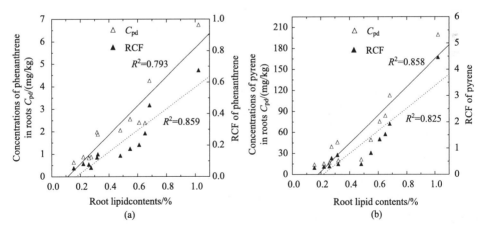

Figure 6-6 Correlations of root lipid content with root concentration or RCFs of phenanthrene and pyrene for tested twelve plant species. (a) phenanthrene, (b) pyrene

0.79 ($n=12$) and 0.86 ($n=12$) (Figure 6-6). These results indicate that root lipid content might be a good predictor for root accumulation of these PAHs. This is, in some way, consistent with the observations on other organic compounds by Chiou et al (2001). They elucidated that for roots with high water contents, the root-water phase acted as the major reservoir for highly water-soluble contaminants. By contrast, the lipids in roots, even at small amounts, were usually the major reservoir for highly water insoluble contaminants.

Root concentration factors (RCFs), defined as the ratio of the PAH concentrations in roots and in soils on a dry weight basis (Briggs et al., 1982, 1983; Polder et al, 1995) were obtained (Table 6-4 and Table 6-5). RCFs generally tended to decrease with the increase of soil PAH concentrations as listed in Table 6-4. However, as to the various plant species grown in the same spiked soils S6 (Table 6-5), RCFs of phenanthrene (0.05~0.67) were much lower than those of pyrene (0.23-4.44) for the same soil-plant treatment. Most importantly, RCFs of phenanthrene and pyrene were significantly positively correlated with root lipid contents at $p < 0.05$ confidence level (Figure 6-6), which should be kept in mind when estimating the potential for human exposure to PAHs from various kind of vegetation. In most cases, RCFs of PAHs shown in Table 6-4 and Table 6-5 were a little larger than those reported by Peterson et al. (2002). They observed that the RCFs of PAHs, including acenaphthylene, fluoranthene, benzo[a]pyrene, and indeno[1,2,3-cd]pyrene, for potato were about 0.002~0.100. Lower soil PAH concentrations and different plant characters such as potato lipid contents probably accounted for the disparities.

In the past, studies have shown that most lipophilic organic compounds (K_{ow} greater than approximately 10^4) partition to the epidermis of the roots (Paterson and Mackey, 1994; Wang and Jones, 1994), and the extent to which a lipophilic organic compound enters plant's roots from contaminated soil depends on the K_{ow}. Generally, the more lipophilicity of organic chemicals results in the higher root concentrations (Trapp et al., 1990). Although there still exists a lack of information on root uptake of PAHs hitherto, results of this work consistently showed that root accumulation of pyrene was greater than phenanthrene as exhibited by root PAH concentrations or RCFs. This would result from the higher K_{ow} of pyrene than that of phenanthrene. Moreover, several studies suggest that the root uptake of lipophilic organic compounds can be in correlation with root compositions such as lipid contents (Simonich and Hites, 1995; Chiou et al., 2001). While, at this time, the compositional effects of plant species or the different parts of a plant have not been sufficiently elucidated, and information was scant on correlations between root concentrations or RCFs of PAHs and root compositions. Results of this study proved that lipid content would be useful in predicting of root accumulation of phenanthrene and pyrene as a consequence of significantly positive correlations of root concentrations and RCFs of these compounds with root lipid contents.

6.2.2 Accumulation of PAHs in shoots

Shoot concentrations of phenanthrene and pyrene as a function of their soil concentrations were shown in Table 6-6. Shoot accumulation of phenanthrene and pyrene also consistently enhanced along with the increase of their soil concentrations. Whereas the concentrations of these compounds in shoots were statistically far lower than roots. Great difference was also displayed between various plant species (Table 6-7). Content of phenanthrene and pyrene in shoots of twelve species grown in S6 varied in range of 0.10~1.96 mg/kg for phenanthrene and 0.23~7.37 mg/kg for pyrene.

Table 6-6 Shoot concentrations (C_{pd}, mg/kg) and concentration factors (SCFs) of phenanthrene and pyrene for flowering Chinese cabbage as a function of their soil concentrations (C_s, mg/kg) after 45 days. Concentrations of phenanthrene and pyrene in soils, roots and shoots were on a dry weight basis

Soil No.	C_s-ph /(mg/kg)	C_s-py /(mg/kg)	Shoots			
			C_{pd}-ph /(mg/kg)	C_{pd}-py /(mg/kg)	SCF-ph/($\times 10^{-1}$)	SCF-py/($\times 10^{-1}$)
S0	ND	ND	0.06(0.01)	0.09(0.01)	-	-
S1	0.85(0.10)	2.71(0.28)	0.14(0.02)	0.22(0.02)	1.70(0.15)	0.81(0.07)
S2	0.93(0.08)	2.81(0.34)	0.17(0.02)	0.29(0.03)	1.79(0.20)	1.03(0.09)
S3	2.81(0.34)	11.3(1.14)	0.22(0.02)	0.46(0.06)	0.77(0.05)	0.41(0.04)
S4	6.71(0.81)	34.0(5.01)	0.24(0.03)	0.86(0.08)	0.35(0.04)	0.25(0.02)
S5	8.60(0.75)	41.9(3.77)	0.25(0.01)	1.07(0.13)	0.28(0.03)	0.26(0.03)
S6	9.42(1.02)	58.8(6.08)	0.30(0.03)	1.28(0.11)	0.31(0.02)	0.22(0.02)
S7	13.8(1.48)	123(14.6)	0.34(0.03)	2.43(0.34)	0.24(0.03)	0.19(0.02)
S8	36.9(4.33)	301(39.2)	0.61(0.06)	4.57(0.37)	0.17(0.01)	0.15(0.01)

Note: C_s-ph and C_s-py were the soil concentrations of phenanthrene and pyrene (mg/kg); C_{pd}-ph and C_{pd}-py were plant concentrations of phenanthrene and pyrene (mg/kg); SCF-ph and SCF-py were shoot concentration factors of phenanthrene and pyrene respectively. ND-not detectable. Data in bracket were standard deviations; $n=3$.

Table 6-7 Shoot concentrations (C_{pd}) and concentration factors (SCFs) of phenanthrene and pyrene in unspiked control soil (S0) and spiked soil with initial phenanthrene concentration of 133 mg/kg and pyrene of 172 mg/kg (S6) after 45 days

Plant No.	S0		S6			
	C_{pd}-ph /(mg/kg)	C_{pd}-py /(mg/kg)	C_{pd}-ph /(mg/kg)	C_{pd}-py /(mg/kg)	SCF-ph ($\times 10^{-1}$)	SCF-py ($\times 10^{-1}$)
P1	0.13(0.01)	0.29(0.02)	1.96(0.22)	7.37(0.63)	1.19(0.28)	1.16(0.12)
P2	0.06(0.05E)	0.09(0.01)	0.30(0.03)	1.28(0.11)	0.31(0.02)	0.22(0.02)
P3	0.05(0.04E)	0.06(0.01)	0.86(0.06)	1.24(0.14)	0.64(0.05)	0.24(0.03)
P4	0.13(0.01)	0.29(0.03)	0.39(0.04)	1.07(0.01)	0.44(0.05)	0.19(0.02)
P5	0.18(0.02E)	0.07(0.01)	0.30(0.02)	0.50(0.04)	0.19(0.02)	0.08(0.01)
P6	0.31(0.03E)	0.88(0.08)	0.56(0.07)	2.09(0.23)	0.56(0.06)	0.47(0.05)
P7	0.26(0.03)	1.05(0.09)	0.35(0.03)	1.53(0.16)	0.24(0.02)	0.25(0.03)
P8	0.15(0.01)	0.26(0.03)	0.78(0.07)	1.94(0.22)	0.64(0.05)	0.34(0.04)
P9	0.14(0.02)	0.25(0.03)	0.76(0.09)	0.90(0.07)	0.69(0.05)	0.16(0.02)
P10	0.04(0.04E)	0.07(0.01)	0.10(0.01)	0.38(0.03)	0.06(0.01)	0.07(0.01)
P11	0.26(0.02)	0.20(0.02)	0.42(0.03)	0.33(0.03)	0.35(0.04)	0.06(0.01)
P12	0.18(0.02)	0.10(0.01)	0.23(0.02)	0.23(0.01)	0.21(0.02)	0.04(0.01)

Note: C_{pd}-ph and C_{pd}-ph were shoot concentrations of phenanthrene and pyrene; SCF-ph and SCF-py were shoot concentration factors of phenanthrene and pyrene respectively. E meant $\times 10^{-1}$. Data in bracket were standard deviations; $n=3$.

Shoot concentration factors (SCFs), defined as the ratio of PAH concentrations in shoot and in soil on a dry weight basis, were calculated. SCFs of phenanthrene and pyrene generally tended to decrease with the increase of their soil concentrations (Table 6-6). As to the twelve plant species grown in S6, SCFs were 0.006~0.12 for phenanthrene and 0.004~0.12 for pyrene (Table 6-7), which were much lower than RCFs for the same soil-plant treatment. This indicated that the translocation of test PAHs from root to shoot was considerably restricted. Further statistically analysis was conducted, and no significant correlation was found between shoot concentrations or SCFs of phenanthrene and pyrene and shoot compositions.

It was notable that even did shoots grown in the unspiked control soil, shoot accumulation of phenanthrene and pyrene was obviously detectable (Table 6-7). The respective concentrations of phenanthrene and pyrene in twelve plant shoots grown in S0 amounted to 0.04~0.31 mg/kg for phenanthrene and 0.06~1.05 mg/kg for pyrene, which should only derive of shoot uptake and accumulation from atmosphere probably through the retention of vapor phase of PAHs on the waxy leaf cuticle (Bacci et al., 1990; Trapp et al., 1990; Polder et al., 1995). It indicates that the shoot uptake of PAHs from the ambient air, possibly originally volatized from the soils, was an important pathway for these PAH intake by vegetable above-ground part. Then phenanthrene and pyrene in shoots grown in S6 should be approximately the sum of shoot accumulation from air (denoted as the concentrations of phenanthrene and pyrene in plant shoots grown in S0) and translocation from roots (Schroll et al., 1994). This was common for foliage uptake of semivolatile organic chemicals in field conditions.

Although air concentrations of PAHs had not been controlled through the entire experiment procedure, which was similar to the field conditions, it would be reasonable to assume that the air conditions were homogeneous for all the soil-plant treatments. Because the various treated pots were randomized in the small glass greenhouse side by side and exchanged places frequently, and the PAH concentrations in the air would quickly become homogeneous after volatilizing from soils (Ryan et al., 1988; Wang and Jones, 1994; Schroll et al., 1994; Trapp and Matthies, 1997). This was also supported by the results of the identical monitored air PAH concentrations in the greenhouse 5 cm or 20 cm above different soil surface. As a result, the disparities of shoot PAH concentrations for various plant species grown in S0 would come from plant properties. It has been suggested that shoot uptake and accumulation of lipophilic organic compounds from atmosphere, in part, depend on the plant lipid contents and plant surface area (Simonich and Hites, 1995; Kipopoulou et al., 1999; Howsam et al., 2001). However, results of our study revealed that R^2 values of linear regression for correlations of shoot lipid contents and shoot PAH concentrations in S0 were relatively too low to draw positive conclusions (R^2 for phenanthrene and pyrene were 0.6214 and 0.5207, respectively; $n=12$).

6.2.3 Translocation of PAHs in plant

Although shoots apparently accumulated phenanthrene and pyrene from air, the concentrations of phenanthrene or pyrene in shoots grown in spiked soils were still much larger than shoot uptake and accumulation of these compounds from atmosphere (Table 6-4 and Table 6-7). This in fact suggests that despite of shoot uptake and accumulation from air, the translocation of phenanthrene or pyrene from roots to shoots was also significant. According to the calculation methods of Schroll et al. (1994) for agricultural plants uptake pathway of organic chemicals from soils, results showed that about 22%~95% of phenanthrene and 32%~96% of pyrene in shoots were translocated from root uptake. On the other hand, roots of twelve plant species grown in unspiked control soils were also analyzed. Phenanthrene or pyrene in these roots was not detectable, implying that phenanthrene and pyrene accumulated by various shoots from atmosphere were rarely translocated to the roots.

It has been thought that lipophilic organic pollutants partition to the epidermis of root and are not drawn into the inner root or xylem (Kipopoulou et al., 1999). Based on this assumption, studies have focused on the foliage uptake and accumulation of PAHs, and information was scarce on the root uptake and subsequent translocation of these chemicals (Trapp et al., 1990; Simonich and Hites, 1995; Kipopoulou et al., 1999). As a consequence, there are difficulties to evaluate the translocation of PAHs from roots. Results of this work showed that although K_{ow} of phenanthrene or pyrene is relatively high (lg K_{ow} exhibits 4.57 for phenanthrene and 5.18 for pyrene) (Polder et al., 1995), the translocation of these compounds from roots to shoots was positive, and was usually the major pathway of shoot accumulation of these compounds. However, the translocation of phenanthrene and pyrene in plant tissues was speculated on approximate mass balance of root or shoot accumulation of these compounds. In this research, there were no ways to estimate the xylem or phloem flow of PAHs necessary to transport the observed amounts in root-shoot system taking into account the solubility of PAHs in the aqueous xylem or phloem sap and see if this corresponded to realistic evapor transpiration rates, which should be deepened in future.

6.3 Comparison for plant uptake of PAHs from soil and water

Uptake by plants from the surrounding soil and water is an important step for the transfer of organic toxic contaminants into the food chain/web. It is notable that, different from the plant uptake of lipophilic organic chemicals from aqueous solution, the availability of these chemicals in soils has been shown to be a primary restriction for their effective uptake by plants from soils (Gao et al., 2005). As highly nonpolar hydrophobic molecules, PAHs tend to strongly partition into soil solids (particularly the soil organic fractions) (Gao et al., 2006; 2007; 2010b). The sorption/ desorption process of PAHs in soil would affect their uptake by plants, which, in theory, leads to some difference between plant uptake from soil and aqueous solution.

In this section, we seek to comparably evaluate the plant uptake of PAHs, phenanthrene and pyrene as representatives, from soil and water using greenhouse experiments (Gao and Ling, 2006). Ryegrass (*Lolium multiflorum* Lam.) was used as the uptake plant throughout the experiments. Plant uptake of phenanthrene and pyrene in aqueous solution was conducted using a batch technique (Li et al., 2002). Ryegrass seeds were germinated and seedlings were cultivated. Then seedlings were transplanted into Hoagland solutions with 1.00 mg/L phenanthrene and 0.12 mg/L pyrene. Plants were sampled and detected for PAH concentration after 0~12 days cultivation. Soil pot experiments were conducted to investigate the uptake of PAHs from soil. Soils were collected from an A (0~20 cm) horizon previously free of PAHs, with pH 5.05 and 14.5 g/kg organic matter. Soils were sieved, PAH-spiked, aged and potted for plant vegetation. Pregerminated seeds of ryegrass were sown, and seedlings were managed. Soils and plants were destructively sampled after 45 days since sowing. Plants were washed with distilled water, dried with filter paper, and detected for PAH concentrations (Gao and Ling, 2006).

6.3.1 Plant uptake of PAHs from water

Plant uptake of phenanthrene and pyrene from aqueous solution as a function of time is shown in Figure 6-7. Root and shoot concentrations of PAHs were on a dry weight basis. The original fast uptake of these compounds occurred and resulted in a sharp rise in the root and shoot concentrations of ryegrass and a large drop in the corresponding aqueous concentrations (Figure 6-8). After about 48 h, the concentrations of phenanthrene and pyrene reached maximums of 104.7 mg/kg and 31.41 mg/kg, respectively, in ryegrass roots, and 2.86 mg/kg and 0.303 mg/kg, respectively, in ryegrass shoots. Thereafter, the concentrations of phenanthrene and pyrene in both plants and aqueous solutions decreased gradually with uptake time.

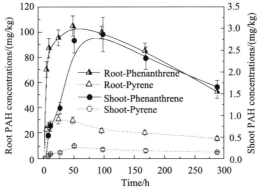

Figure 6-7 Plant concentrations of phenanthrene and pyrene as a function of uptake time. Error bars represent ±1 SD

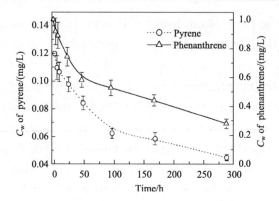

Figure 6-8 Concentrations of phenanthrene and pyrene in culture solution as a function of uptake time. Error bars represent ±1 SD

It has been well documented that shoot concentrations of organic chemicals might be greatly different from the roots concentrations. For hydrophobic compounds, on the premise that the soil-air-shoot uptake pathway does not dominate, shoot concentrations were generally much lower than root concentrations (Wang and Jones, 1994; Mattina et al., 2003). In this investigation, although shoot uptake of phenanthrene and pyrene was similar to the tendency of root uptake of these compounds (Figure 6-7), it is obvious that shoot concentrations of test PAHs were always much lower than root concentrations over the entire experimental duration. In addition, the measured concentrations of phenanthrene and pyrene in ryegrass shoots from unpolluted control nutrient solution were only 0.081 mg/kg and 0.021 mg/kg at 288 h, which were negligible compared to those grown in polluted conditions. This indicates that the PAHs in shoots resulted primarily from translocation from roots.

The variation of plant concentrations of phenanthrene and pyrene as a function of time, in principle, depended on the water concentrations, plant growth, uptake, and metabolism rate of these chemicals. It should be noted that significant exponent curves were shown for root or shoot biomass of ryegrass as a function of uptake time from 0 h to 288 h. In the duration of the initial stage of uptake, root or shoot biomass increased a little. Thereafter, the roots and shoots enlarged sharply (data not shown). It was reported that the metabolism of such chemicals is insignificant (Trapp and Matthies, 1995). Thus, the apparent decrease of plant PAH concentration after the initial increasing stage might be primarily ascribed to the great decrease of PAH solution concentration and the fast plant growth dilution.

6.3.2 Plant uptake of PAHs from soil

The ryegrass uptake of phenanthrene and pyrene from soils as a function of their soil concentrations after 45 days was displayed in Figure 6-9. Plant concentrations were on a dry weight basis. As seen, root concentrations of phenanthrene and pyrene monotonically enhanced with the increasing soil concentrations. Along with the increments of soil phenanthrene and pyrene concentrations from 0.646 mg/kg to 43.50 mg/kg and from 0.864 mg/kg to 272.1 mg/kg, phenanthrene and pyrene concentrations in roots of ryegrass increased from 0.137 mg/kg to 4.68 mg/kg and from 0.456 mg/kg to 42.5 mg/kg, respectively. Because concentrations of test PAHs in roots grown in unspiked control soils were not detectable, it could be concluded that these PAHs in roots derived directly from root uptake from the surrounding soils. In addition, root concentrations of pyrene were generally much greater than those of phenanthrene at the same soil concentrations.

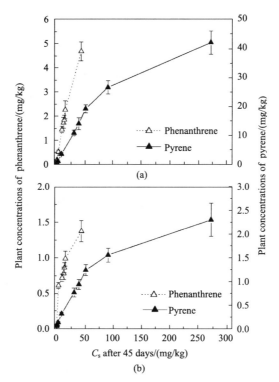

Figure 6-9 Concentrations of phenanthrene and pyrene in roots and shoots of ryegrass as a function of PAH concentrations in soils after 45 days. (a) Root. (b) Shoot. Error bars represent ±1 SD

It has been viewed that organic chemicals with lg K_{ow} values of about 4.5 were most likely to be translocated from roots to shoots and accumulated in the stem and leaf tissues of plants (Briggs et al., 1982; Ryan et al., 1988). Lg K_{ow} values of phenanthrene and pyrene are about equal to or a little larger than 4.5, so translocation would be significant. As shown in Figure 6-9, concentrations of phenanthrene and pyrene in the shoots of ryegrass also monotonically increased with the increment of their soil concentrations. However, when comparing with roots, PAH concentrations in shoots were remarkably lower, implying that the transport of phenanthrene and pyrene from roots to shoots was markedly restricted. One notes that even for the unspiked control soils, shoot concentrations of test PAHs were detectable. Concentrations of phenanthrene and pyrene in shoots of ryegrass grown in the unspiked control soils after 45 days were 0.047 mg/kg and 0.060 mg/kg, respectively, which were much lower than those in PAH-spiked soil treatments (Figure 6-9). These results indicate that shoot uptake of these compounds from the air might be a minor pathway of shoot accumulation, as stated previously.

6.3.3 Comparison for plant uptake of PAHs from soil and water

To compare the PAH uptake by plants from soil and water, the PAH solution concentration (C_w) in soil interstitial water was obtained as follows

$$C_w = C_s / (f_{som} K_{som})$$

where C_s is the concentration of PAHs in soil on a dry weight basis. f_{som} is the weight fraction of the soil organic matter (SOM) in soil. Here, f_{som} is 1.45%. K_{som} is the contaminant partition coefficient between SOM and water (Chiou et al., 2001). The K_{som} values of phenanthrene and pyrene, estimated from the correlation of lg K_{som} = 0.906 lg K_{ow}−0.779 (Chiou et al., 1983), were 1790 and 4291, respectively. Plant concentration factors (PCFs) defined as the ratio of contaminant concentration in plants on a dry weight basis (C_p) to that in culture solution or in soil interstitial water (C_w), were calculated as follows

$$PCF = C_p / C_w$$

Then plant concentrations and PCFs as a function of the PAH solution concentration (C_w) in culture solution or in soil interstitial water could be obtained and compared.

The ryegrass uptake of phenanthrene as a representative PAH from soil and water as a function of its concentrations in culture solution or soil interstitial water (C_w) was given in Figure 6-10. It has been observed that RCFs increased with increasing K_{ow}, and the sorption of chemicals by macerated roots was very closely related to the RCFs of the living root for lipophilic chemicals with lg K_{ow} values larger than 3 (Briggs et al., 1983). As a lipophilic organic chemical, the crucial step for root uptake of PAH was its sorption onto the root surface. As seen from Figure 6-10, plant concentrations of phenanthrene generally increased with the increasing phenanthrene concentrations in solution or soil pore water (C_w) irrespective of the soil-plant and water-plant systems. Similarly, as discussed above, root concentrations and RCFs were generally much larger than shoot concentrations and SCFs of phenanthrene for ryegrass uptake from soil or water.

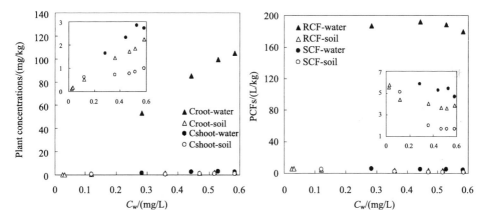

Figure 6-10 Plant concentrations and PCFs of phenanthrene as a function of phenanthrene concentrations in culture solution or soil interstitial water (C_w). Cplant-water and Cplant-soil are concentrations of phenanthrene in ryegrass grown in culture solution and soil, respectively. PCF-water and PCF-soil are PCFs for ryegrass uptake of phenanthrene from culture solution and soil, respectively

A novel finding was observed that plant concentrations and PCFs of phenanthrene for ryegrass uptake from culture solution were always much higher than those for ryegrass uptake from soils at the same C_w, irrespective of root and shoot. A similar tendency was also observed for plant uptake of pyrene from the surrounding environment (data not shown). It has generally been accepted that organics in soil interstitial bulk-like water would be available for plant-root uptake, followed by translocation from root to shoot (Chiou et al., 2001). But as for highly hydrophobic organic chemicals such as PAHs, they tend to strongly partition into soil solids, particularly the soil organic fractions. The desorption process of these compounds from soil solids into soil pore water restricts their uptake by plants, which would result in lower plant concentrations and PCFs of test PAHs in soil-plant systems than those in water-plant ones. These results suggest that PAHs in culture solution would be more available and susceptible to uptake by plants than those in soil interstitial water.

Chapter 7   Subcellular distribution of PAHs in plants

Understanding of how plants take up and accumulate PAHs from the soil environment could have considerable benefit for risk assessments of PAH contamination (Aajoud et al., 2006; White et al., 2005). In general, two primary processes are responsible for PAH transfer and distribution in plant tissues: ① transfer between plant tissues and cells driven by transpiration and the PAH concentration gradient across plant-cell components, and ② accumulation of PAHs in plant tissues, with the extent related to plant lipid contents (Paterson and Mackay, 1994; Wang and Jones, 1994; Gao and Zhu, 2004; Gao and Collins, 2009). However, the PAH transfer and distribution in plant intracellular tissues are still unclear. In this chapter, the subcellular distribution of PAHs in plants was elucidated. Results will enhance the understanding of PAH transfer mechanisms in inner plants and their effects on the subcellular distribution of PAHs in plants.

7.1   PAH distribution in subcellular root tissues

Uptake from water and soil via plant roots is a major pathway of PAH entry into plants. Wild et al. (2005) reported that PAHs first adsorbed to root surfaces and then passed through the membranes of adjoining cells before accumulating in cell walls and vacuoles. The amount of uptake depended primarily on the lipid content of plant roots, in which protein, fats, nucleic acids, cellulose tissues, and other components all contain lipophilic components, which appear to be the primary domains where PAHs accumulate once they penetrate plant root cells (Gao and Zhu, 2004). Unfortunately, despite extensive studies on the transport of organic contaminants (especially PAHs) in plants, information about PAH distributions in intracellular tissues of plant roots, stalks, and leaves is lacking.

Here, we investigated the uptake and subcellular distributions of PAHs in root cells of ryegrass (*Lolium multiflorum* Lam.), which is widely used in the rhizoremediation of PAH-contaminated sites owing to its fibrous root system and large specific root surface area. Ryegrass root uptake of phenanthrene and pyrene as representative PAHs from aqueous Hoagland solution was investigated using batch settings in a greenhouse (Kang et al., 2010). Following germination in vermiculite, ryegrass seedlings were transferred in Hoagland solution and grown for 2 weeks. Then seedlings were transplanted to Hoagland culture solutions containing phenanthrene and pyrene. At 0~240 h exposure, the seedlings were sampled and prepared for PAH analysis.

7.1.1 Fractionation protocol of root subcellular tissues

A modification of the methods of Lai et al. (2006), Li et al. (2006b), and Wei et al. (2005) was used to obtain subcellular fractions of root cells. Briefly, fresh roots were mixed with extraction buffer containing 50 mmol/L HEPES, 500 mmol/L sucrose, 1.0 mmol/L DTT, 5.0 mmol/L ascorbic acid, and 1.0% (w/v) polyclar AT PVPP. The buffer solution pH was adjusted to 7.5 using 1.0 mol/L NaOH. The root tissue extract was ground, passed through a 60 μm sieve, and subsequently centrifuged at 500 g for 5 min to obtain a pellet of cell debris. This pellet was referred to as the wall fraction of the root cells. The supernatant was then centrifuged at 10 000 g for 30 min to obtain the cell organelle fraction. All extraction steps were performed at 4℃. The dried cell-wall and organelle powders were placed in a vacuum freeze drier (Labconco, Kansas City, MO, USA) at –65℃. Fraction contents, determined gravimetrically, were 6.0% organelles and 8.9% cell walls, with the remainder being water and water-soluble fractions.

7.1.2 Uptake of PAHs by roots

The uptake of phenanthrene and pyrene from Hoagland medium by ryegrass roots as a function of exposure time is shown in Figure 7-1. The uptake rate and magnitude of uptake of phenanthrene and pyrene by ryegrass roots differed. Concentrations of phenanthrene and pyrene in roots increased with exposure time, reaching a maximum at about 100 h. Although phenanthrene concentrations in roots were higher within this timeframe, they were less than two times the concentration of pyrene, most likely because of a higher initial concentration in the medium of phenanthrene (2.5 mg/L) than of pyrene (0.5 mg/L). From 100 h to 240 h, the phenanthrene concentration in roots decreased sharply from 90 mg/kg to 18 mg/kg, whereas that of pyrene declined gradually from 60 mg/kg to 48 mg/kg. These differing uptake patterns could result from a difference in the migration of phenanthrene and pyrene to ryegrass shoots, their degradation in roots, or both. The slower rate of pyrene reduction after 100 h indicates that pyrene is recalcitrant to metabolism in roots and that, as the more lipophilic compound, it exhibits a strong affinity for plant tissues, slowing its transport from roots to shoots.

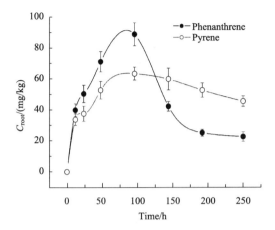

Figure 7-1　Concentrations of phenanthrene and pyrene in ryegrass roots as a function of uptake time. C_{root} means concentrations of phenanthrene and pyrene in ryegrass root

Although spontaneous volatilization could be a cause of PAH dissipation from water, it is thought to be primarily related to plant accumulation and metabolism (Gao et al., 2006b). Figure 7-2 shows the dissipation efficiency of PAHs from Hoagland solution by ryegrass, which we defined as the ratio of PAH removal to the initial concentration of PAH in aqueous solution. Dissipation efficiency increased gradually with exposure time, reaching 92% for phenanthrene and 62% for pyrene at 250 h. This result is consistent with the relatively rapid reduction of phenanthrene in ryegrass roots, which could be due to its high metabolism rate in ryegrass, relatively quick migration from roots to shoots, or both.

The root concentration factor (RCF) describes the capability of roots to accumulate contaminants from direct contact with an aqueous environment, which is here defined as the ratio of PAH concentration in roots (C_{root}) to that in the culture medium ($C_{solution}$): RCF = $C_{root}/C_{solution}$ (Briggs et al., 1982; 1983; Polder et al., 1995). RCF values increased with increasing root/solution contact time before approaching a nearly constant value after 150 h (Figure 7-3). The RCF values of pyrene were about two times greater than those of phenanthrene. Our previous study indicated that the lipophilicity (e.g., lg K_{ow}) of a compound is a determinant of the magnitude of plant uptake [18]. The higher pyrene RCF is due to its greater lipophilicity (lg K_{ow} = 5.32) compared with that of phenanthrene (lg K_{ow} = 4.46). Gao and Zhu (2004) reported that

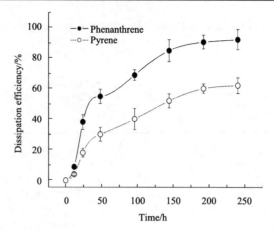

Figure 7-2 Dissipation efficiency (%) of phenanthrene and pyrene by ryegrass from aqueous solution as a function of exposure time. The values were defined as ratio of PAH removal to the initial concentration from the aqueous solution. It explained in principle the plant-affected dissipation of phenanthrene and pyrene from solution, and such dissipation was primarily related to plant accumulation and metabolism

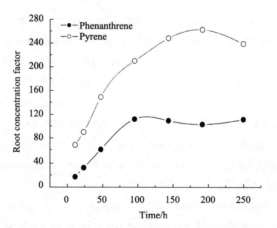

Figure 7-3 Root concentration factor (RCF) of phenanthrene and pyrene for ryegrass uptake from aqueous solution. Root concentration factor (RCF) describes the capability of roots to accumulate contaminants from the direct contact with the aqueous environment, which is here defined as the ratio of PAH concentration in root (C_{root}) to that in culture medium ($C_{solution}$), i.e., RCF = C_{root} / $C_{solution}$

pyrene uptake by plants from soil was 4~7 times greater than uptake of phenanthrene. Together, these results suggest that more lipophilic organic contaminants have a higher propensity for uptake in plants via roots.

7.1.3 Subcellular movement and distribution of PAHs in root cells

Phenanthrene and pyrene concentrations in root subcellular fractions were displayed in Figure 7-4. Phenanthrene concentrations in cell walls and organelles increased gradually to 79 mg/kg and 95 mg/kg, respectively, and then decreased to 16.5 mg/kg and 17 mg/kg as exposure time lengthened. Pyrene underwent a similar uptake pattern: pyrene concentrations in cell walls and organelles rose to 58 mg/kg and 71 mg/kg, respectively, and then decreased to 38 mg/kg and 56 mg/kg (Figure 7-4). Before 70 h of uptake, concentrations of both PAHs in cell walls were greater than those in organelles. Moreover, the uptake by organelle components was slower in reaching a maximum relative to cell walls. The two PAHs first adsorbed onto cell walls from the culture medium, and they then diffused into cell organelle components. After 70 h, a relatively higher concentration of both PAHs was found in the organelle fraction than in cell walls due to the greater accumulation of lipophilic compounds in the fraction containing a higher lipid content, i.e., organelle components. After 96 h of exposure, phenanthrene uptake rapidly decreased in root cell walls and organelles, whereas the decrease in pyrene was much slower in these two subcellular fractions. This trend is consistent with the uptake patterns of the two contaminants by ryegrass roots shown in Figure 7-1.

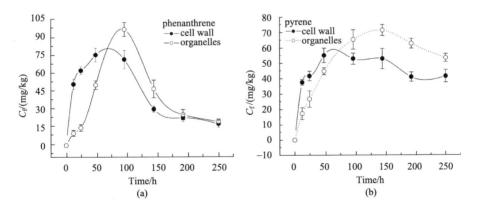

Figure 7-4 Concentrations of phenanthrene and pyrene in root cell walls and organelles as a function of time. C_f means concentrations of phenanthrene and pyrene in root subcellular fractions

PAHs in roots were distributed into three subcellular fractions: water-soluble fraction, cell wall, and organelle. The proportions of phenanthrene and pyrene in the water-soluble, wall, and organelle fractions are plotted against exposure time in

Figure 7-5. From 12 h to 240 h of exposure, the percentage of phenanthrene in cell walls notably descended from 84% to 42% and pyrene decreased from 60% to 41%. Within the same period, phenanthrene and pyrene distributions in organelles increased from 8.5% to 41% and from 21% to 33%, respectively. Both PAHs showed a relatively small variation in the proportion in the aqueous soluble fraction, about 10%~15% for phenanthrene and 10%~20% for pyrene. Thus, at the beginning of uptake (i.e., < 96 h), the decrease of PAH in cell walls largely corresponded to the increase in cell organelles, suggesting that PAHs first accumulated in cell walls via direct contact with Hoagland solution and then gradually transferred to fractions inside cells, such as organelles. After 96 h of exposure, the distributions of both PAHs in cell components approached a relatively stable state in cell walls and organelles (Figure 7-5). In ryegrass root cells, content in cell walls was 8.9%, and that in organelles was 6.0%. The relatively smaller organelle fraction accumulated a similar amount of each PAH compared with that in cell walls, likely owing to the higher lipid content of organelles. Generally, the uptake capability of root tissues for organic lipophilic compounds increases with K_{ow} value (i.e., $K_{ow} > \sim 10^4$), with more lipophilic compounds showing a higher accumulation in plants, particularly in plant tissues containing a high lipid content (Paterson and Mackay, 1994; Wang and Jones, 1994; Gao et al., 2005). Lipids in plant cell walls are composed mostly of polysaccharides (90%), with a few structural proteins, lignin, lectin, and mineral elements as well as a very small lipid component. In contrast, the lipid content of plant organelles is 15%~30% (Li, 2006), enabling them to draw PAHs from the cell wall. Thus, the relatively higher lipid content of the organelle fraction is believed to be responsible for the greater accumulation of PAHs, and the corresponding concentration gradient established between organelles and cell walls in the beginning stage of uptake (< 96 h) is the driving force for the diffusion of PAHs to interior cell components.

As shown in Figure 7-5, PAH concentrations in soluble components stayed nearly constant. The separated soluble cellular components, mainly consisting of cell solution and largely concentrated in the cell matrix between cells or organelles, can be regarded as an intracellular buffering distribution phase. Due to the hydrophobicity of PAHs, these aqueous substances were not easily enriched. Thus, the non-affinity between PAHs and soluble cellular components may result in distributive constant and low partitioning proportions.

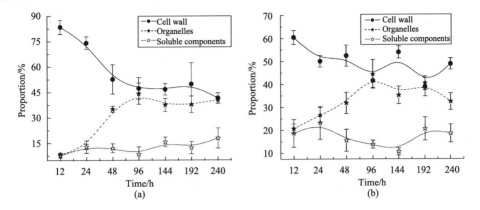

Figure 7-5 Proportions of phenanthrene (a) and pyrene (b) distributed in cell water soluble fraction, wall and organelle as a function of uptake time. The proportion of PAHs was calculated from the measurement of PAH in each fraction to the total amount in ryegrass root cells

Subcellular fraction-concentration factor (SFCF) values, defined as the ratio of PAH concentration in subcellular fractions to that in water-soluble cellular components, are shown in Figure 7-6. The SFCF of phenanthrene in cell walls decreased from 101 L/kg to 20 L/kg over 240 h of exposure. For pyrene, the SFCF in cell walls first increased from 16 L/kg to 48 L/kg and then decreased rapidly to < 25 L/kg. The difference in cell-wall SFCFs of phenanthrene and pyrene likely resulted from the different properties of the two PAHs, as in the beginning stage, pyrene tended to accumulate more in cell walls than in water-soluble components owing to its higher lg K_{ow}.

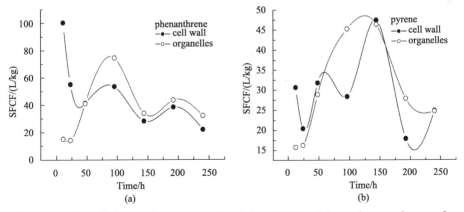

Figure 7-6 Subcellular fraction concentration factors (SFCF) of phenanthrene and pyrene for ryegrass root uptake as a function of time 0~240 h. SFCF was defined as the ratio of PAH concentration in subcellular fractions including cell wall and organelle to that in cell water-soluble components

In the initial 48 h, the SFCFs of the two PAHs were greater in cell walls than in cell organelles. After that period, organelle fraction SFCFs slightly exceeded those of cell walls. These results suggest that within the first 48 h of exposure, subcellular transport of PAHs occurred from cell walls to intracellular organelles as a result of the concentration gradient.

In summary, to our best knowledge, this is the first paper reporting the distribution of persistent organic pollutants, with PAHs as representatives, in plant subcellular tissues. Transpiration is generally considered to be the main transfer mechanism of PAHs in plants, such as from roots to stalks and leaves. PAHs initially adsorb to plant cell walls and then gradually diffuse into subcellular tissues. The lipid contents of intracellular components determine the extent of lipophilic compound accumulation, and the diffusion rate is related to the concentration gradient established between cell walls and organelles inside cells. Although both phenanthrene and pyrene are grouped among organic compounds that share similar properties, pyrene displays greater accumulation factors in subcellular walls and organelle fractions due to a higher lg K_{ow}. One notes that, although the negligible amounts of pure cell membrane could not be separated from other cell fractions by the centrifugal method and it was merged into the soluble components in the investigation, results of this work open new insights into POP subcellular transport and distribution in plants.

7.2 Subcellular distribution of PAHs in arbuscular mycorrhizal roots

Arbuscular mycorrhizal fungi (AMF) form symbioses with plants and are ubiquitous in the environment. The impacts of AMF inoculation on the uptake of organic pollutants by plants have recently been reported. Cheng et al. (2008) and Huang et al. (2007) found that AMF colonization led to increased accumulation of PAHs (pyrene) and DDT in roots, but a decrease in shoots of maize (*Zea mays* L.) and clover (*Trifolium subterraneum* L.). Later, we proved that AMF hyphae play major roles in the uptake of organic chemicals from soils (Gao et al., 2010c). AMF hyphae extend into PAH-contaminated soils and clearly absorb and transport PAHs to roots, resulting in high concentrations of fluorene and phenanthrene in roots (Gao et al., 2010c). These findings highlight the importance of AMF inoculation on plant uptake of organic contaminants. However, the subcellular process and distribution of PAHs in arbuscular mycorrhizal plants still remains to be elucidated.

To this end, the present work aimed to investigate the subcellular distribution of PAH, acenaphthene (ACE) as a representative, in arbuscular mycorrhizal roots of ryegrass (*Lolium multiflorum* Lam.) (Gao et al., 2011b). Two AMFs, *Glomus mosseae* (AMF1) and *G. etunicatum* (AMF2), were experimented. A soil (Typic Paleudalf) was collected from the A horizon (0~20 cm) in Nanjing City, China. The soil had a pH of 6.02, organic carbon content of 14.3 g/kg, clay of 24.7%, sand of 13.4%, and silt of 61.9%. PAH-contaminated soils were obtained according to Ling et al (2010). The final measured concentrations of ACE in soil 1 (S1) and soil 2 (S2) were 18.99 mg/kg and 44.46 mg/kg, respectively. Seeds of ryegrass were sown in a greenhouse in treated soils after AMF inoculation (Gao et al., 2010c). Plants were sampled after 40 days and 70 days. The subcellular fractionation protocol was stated in "7.1.1". The content of each root fraction, determined gravimetrically, was 5.98% organelles, 8.92% cell walls, and the remaining (85.1%) was cell solutions (cell water and water-soluble fraction). ACE in the cell wall, organelles, and solution was extracted and determined according to Kang et al (2010).

7.2.1 PAH concentrations in subcellular tissues of arbuscular mycorrhizal roots

The measured concentrations of ACE in the cell wall and organelles of arbuscular mycorrhizal roots of ryegrass are shown in Table 7-1. ACE concentrations in the AMF root organelles in S1 and S2 after 40~70 days were 18.7~42.5 mg/kg (on a dry weight basis), which was 58%~437% higher than ACE in cell walls. A similar trend was also observed in non-AMF control treatments (CKs). These results clearly prove that ACE passed through the cell wall boundary and entered cell organelles of root cells.

Table 7-1 Concentrations of ACE in cell wall and organelles of arbuscular mycorrhizal root of ryegrass (mg/kg, on a dry weight basis)

Treatment	Subcellular fraction	70 days		40 days
		S1	S2	S2
AMF1	Cell wall	10.3 (0.90)	7.87 (2.99)	7.91 (0.79)
	Organelle	38.4 (9.56)	37.6 (5.48)	42.5 (21.6)
AMF2	Cell wall	9.75 (1.83)	6.15 (1.92)	21.5 (5.28)
	Organelle	18.7 (5.13)	27.7 (8.70)	33.9 (14.1)
CK	Cell wall	11.6 (7.10)	12.9 (4.14)	15.2 (7.60)
	Organelle	40.3 (12.7)	48.3 (10.8)	26.9 (1.86)

Note: Data in bracket are standard deviation (SD).

As reported (Gao and Collins, 2009), plant uptake of PAHs is necessarily accompanied by water from the transpiration stream, and water moves into root systems through apoplastic and symplastic pathways (Figure 7-7). Apoplastic water movement involves diffusion between cell walls, rather than entry into cells, whereas symplastic water movement is through the cell cytoplasm or vacuole and to interconnected cells via plasmodesmata (Wild et al., 2005). ACE in water penetrated into the inner cells of roots along these two pathways, although some ACE was captured by the cell walls.

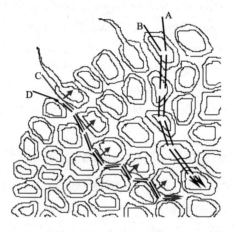

Figure 7-7　Pathways of water and ACE movement across the root epidermis and cortex.

A: Symplastic ACE movement; B: Symplastic water movement; C: Apoplastic ACE movement;

D: Apoplastic water movement

The determined concentrations of ACE in the cell solution ($C_{solution}$, mg/L) further proved the above observations. $C_{solution} = M_{solution}/V_{solution}$, where $M_{solution}$ (mg) was the detected amount of ACE in cell solution per unit root, and $V_{solution}$ (L) was the volume of cell solution per unit root. Here, the $C_{solution}$ of ACE in the AMF1-inoculated roots, AMF2-inoculated roots, and control roots after 40 days were 0.403 mg/L, 1.199 mg/L and 0.518 mg/L, respectively. This means that ACE penetrated into inner cells of the root, and dissolved in cell solution.

The detected high concentrations of ACE in organelles derived from ACE partitioning from the cell solution. As known, hydrophobic organic chemicals with low solubility like PAHs are readily partitioned into the solid phase in a solid-solution system (Li et al., 2005). The partitioning ability depends on the chemical characteristics, solid composition, and solution conditions. Lipophilic organic chemicals are more readily partitioned into solids (roots), which have higher lipid content (Collins et al., 2006). In this case, ACE as a lipophilic chemical, was preferentially located in organelles that have lipophilic components (Kang et al., 2010), resulting in high concentrations in this subcellular fraction.

7.2.2 Subcellular concentration factors of PAH in arbuscular mycorrhizal roots

We further calculated the subcellular concentration factors (SCF, L/kg) of ACE for the cell wall and organelles. SCF = $C_{solid}/C_{solution}$, where C_{solid} (mg/kg) is the ACE concentration in the subcellular fraction (on a dry weight basis). As shown in Figure 7-8, the SCFs of ACE for cell organelles in AMF1-inoculated roots, AMF2-inoculated roots, and control roots were 105.4 L/kg, 28.28 L/kg and 51.91 L/kg, respectively, which were significantly higher than those in the cell walls (19.63 L/kg, 17.94 L/kg and 29.33 L/kg, respectively). This indicates that ACE was more readily partitioned into organelles, likely owing to the higher lipid content of organelles. Lipids in plant cell walls are composed mostly of polysaccharides (90%), with a few structural proteins, lignin, lectin, and mineral elements as well as a very small lipid component (Kang et al., 2010). In contrast, the lipid content of plant organelles is 15%~30% (Li, 2006), enabling organelles to draw ACE from the cell wall. Thus, the relatively higher lipid content of the organelle fraction is believed to be responsible for the greater accumulation of ACE.

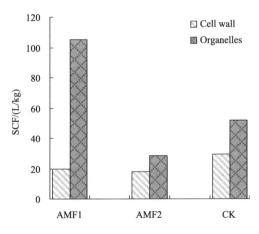

Figure 7-8 The subcellular concentration factors (SCFs) of ACE by cell wall and organelles of arbuscular mycorrhizal root of ryegrass in S2 at 40 days

On the basis of these results, we concluded that the entrance of ACE into the inner cells of AMF-colonized roots of ryegrass could proceed as follows: ACE in apoplastic and symplastic water passes through cell walls, with portions being captured in walls, dissolved in cell solution, and partitioned into cell organelles (Figure 7-7). These results are novel, as the subcellular distribution of PAHs in mycorrhizal plants had not previously been investigated.

7.2.3 Proportion of PAH in subcellular tissues of arbuscular mycorrhizal roots

The proportion of ACE in each subcellular fraction of arbuscular mycorrhizal ryegrass roots was calculated and is shown in Figure 7-9. The cell wall and organelles were the dominant storage compartments for ACE in mycorrhizal roots, with 19.7%~39.8% and 40.8%~70.8% of ACE, respectively, in these two fractions. The cell solution, as the root aqueous phase and distinct from the solid lipid phase, only represents an important storage compartment for organic chemicals with lg K_{ow} < 2 and a dimensionless Henry's law constant of (K_{aw}) <100 (Cousins and Mackay, 2001). Also, only 9.6%~20.5% of ACE is presented in the cell solution. Overall, the distribution of ACE in cells of mycorrhizal ryegrass roots follows, in descending order: cell organelles > cell wall > cell solution.

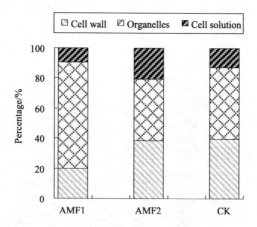

Figure 7-9 Distribution proportions of ACE in each subcellular fraction of arbuscular mycorrhizal root of ryegrass in S2 at 40 days

In summary, this work used a greenhouse experiment to show that, accompanied by the apoplastic and symplastic water movement through the root, ACE passed through the cell-wall boundary, dissolved in the cell solution, and partitioned into organelles in arbuscular mycorrhizal roots of ryegrass. The observed concentrations of ACE in organelles were 0.6~4.4 times higher than in the cell walls. The cell wall and organelles were the dominant storage domains for ACE in the root, and the distribution of ACE in cells of mycorrhizal ryegrass roots was, in descending order: cell organelles (40.8%~70.8%) > cell wall (19.7%~39.8%) > cell solution (9.6%~20.5%).

Chapter 8   Metabolism of PAHs in plants

Following uptake, organic contaminants may be metabolically transformed, thereby reducing their concentration within plant tissues. Plant metabolism has been reported for a relatively small number of industrial organic chemicals, although more is known about herbicides (Van Eerd et al., 2003; VandenHeuvel et al., 1996). Chemicals reported to have been the most successfully metabolized include trichloroethylene, benzene, and explosives. (Shang et al., 2001; Collins et al., 2006; Wild et al., 2006). Metabolic processes and rates are specific to particular chemicals and plant species. However, relatively little is known about the metabolism of persistent organic pollutants such as PAHs in plants. A recent study showed that phenanthrene was metabolized into other polar products in *Zea mays* (Wild et al., 2006). In another study, anthracene was metabolized primarily in cell walls, and the formed products were bound to cell wall components such as pectin, lignin, hemicellulose, and cellulose (harms, 1996; Wild et al., 2005). To our knowledge, only a few PAHs have been previously reported in regard to plant metabolism (Harms, 1996; Li et al., 2001; Brady et al., 2003; Wild et al., 2005), and the involved mechanism is still under consideration.

In this chapter, the metabolism of PAHs in plants was elucidated. Results suggest the great role of enzymes in PAH metabolism in inner plants. The response and distribution of several enzymes in PAH-contaminated plants were clarified. Enzyme inhibitor, ascorbic acid as a representative, may significantly reduce the activities of enzymes thus enhancing the potential risks of PAH contamination in plants. Results of this work will enhance the understanding of PAH transfer and transformation in plants and will be valuable for risk assessments of plant contamination at polluted sites.

8.1   Metabolism of anthracene in tall fescue

Whether persistent organic pollutants are metabolized in plants has been debated in recent decades, and relatively few studies have examined this topic. Plants are capable of transforming polychlorinated biphenyls (PCBs) into non-phytotoxic compounds (Mackova et al., 1997; Chroma et al., 2002a,b). PAHs have long been recognized as potential carcinogens in animals, in which biotransformation into reactive metabolites can lead to DNA damage, and PAH metabolism mainly occurs in the hepatic microsomes (Brady et al., 2003). In recent years, researchers have investigated how plants metabolize PAHs, but the evidence is still scarce, and results are conflicting (Li et al., 2001; Chroma et al., 2002b; Brady et al., 2003; Wild et al., 2005). Using an isotope technique, Li et al. (2001) observed that mineralization and metabolism of phenanthrene were fast in a wheat-solution-lava microcosm, and that the major metabolites of phenanthrene were polar compounds (18% of applied ^{14}C), whereas only 2.1% were identified as non-polar metabolites. The metabolism of the four-ring PAH pyrene has also been investigated using cell suspension cultures of soybean, wheat, purple foxglove, and jimsonweed as well as callus cultures of soybean and foxglove. About 90% of the applied pyrene was transformed in wheat (Hückelhoven et al., 1997). In addition, Brady et al. (2003) provided preliminary evidence for the metabolism of benzo[a]pyrene by *Plantago lanceolata*. On a whole, only a very few documents have been reported on the PAH metabolisms in plants, and limited numbers of PAHs are involved.

Here, our investigation demonstrated the plant metabolism of anthracene(ANT), a typical PAH with three fused benzene rings, based on its reduced concentration, dissipation amount, and detected metabolites in tall fescue (*Festuca arundinacea* Schreb.) (Gao et al., 2013b). First, seedlings were incubated in half-strength Hoagland solution until they reach approximate 15 cm height and relatively mature roots had developed. Then seedlings were transplanted into half-strength Hoagland solution with ANT, and hydroponically cultured for 96 h (The uptake time was based on our previous observed results; Gao et al., 2008b). Then, plants with accumulated ANT were obtained. The treated plants were removed from the PAH-spiked solution, and roots were washed with Milli-Q water (Greenwood et al., 2011; Brady et al., 2003; Li et al., 2005). The plants were then transplanted into half-strength Hoagland culture solution free of ANT. At 0 day, 2 days, 4 days, 8 days and 16 days after the PAH-treated seedlings were transplanted into a clean solution, the seedlings were sampled and prepared for analysis of ANT and its primary metabolites.

8.1.1 Metabolism of anthracene in tall fescue

Tall fescue was first cultured in an ANT-spiked solution for 96 h, and ANT accumulated inside the plants. The treated plants were then transplanted to a culture solution free of ANT. At 0 day, 2 days, 4 days, 8 days and 16 days after transplanting,

ANT concentrations (C_p, mg ANT per kg root or shoot on a dry weight basis) in roots and shoots were examined and data are given in Figure 8-1. C_p in root and shoot at 0 day were 1.204 mg/kg and 0.484 mg/kg, respectively, indicating the clear uptake and accumulation of ANT by tall fescue. However, the ANT concentration clearly decreased from 0 day to 16 days after transplanting. At 16 days, root and shoot concentrations of ANT were only 0.042 mg/kg and 0.055 mg/kg, respectively, which were 96.5% and 88.6% lower than those at 0 day. In 0~16 days after transplanting, the concentrations of ANT in culture solution were undetectable (the detection limit of ANT in aqueous solution was 0.01 mg/L). The temporal change in plant concentrations of ANT theoretically depended on plant growth dilution, phytovolatilization, or metabolism. Root and shoot biomass increased <12.2% in 16 days, suggesting that plant growth dilution was not a dominant contribution to the decrease in plant concentrations of ANT. As an organic compound with a relatively high n-octanol-water partition coefficient (K_{ow}) and low Henry's Law constant (H), ANT showed negligible phytovolatilization (Gao and Zhu, 2005). As such, it can be postulated that ANT metabolism in plant bodies dominated the significant decrease in ANT concentrations in roots and shoots.

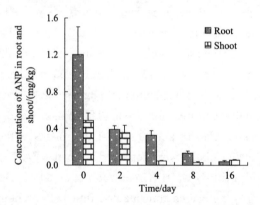

Figure 8-1　Concentrations of anthracene (ANT) in roots and shoots of tall fescue as a function of time

Dissipation of the total amount of ANT in the whole plant, which ignored the influence of plant growth dilution, was examined. Because phytovolatilization of ANT was negligible, the dissipation of ANT in the plants solely depended on plant metabolism. The dissipation ratio (D, %) of the total amount of ANT in the plants was calculated as follows: D (%) = $(M_o - M_i)/M_o \times 100\%$, where M_i and M_o are the total amounts (unit: mg/plant) of ANT in the entire plant at a certain day and day 0, respectively. A larger D value means that a larger portion of ANT was metabolized in the plant. D values as a function of time are given in Figure 8-2. Clearly, D values rapidly increased to 80.2% from 0~4 days and approached an approximate steady state with a D value of approximately 90% at 8~16 days. This result proved that ANT was metabolized in the plants; however, about 10% of ANT remained after 8 days.

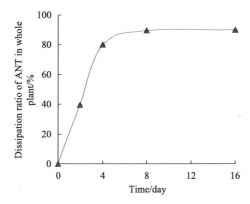

Figure 8-2 Dissipation ratio (D) of anthracene (ANT) in whole tall fescue plants as a function of time. D was calculated as follows: $D\ (\%) = (M_o - M_i)/M_o \times 100\%$, where M_i and M_o are the total amounts of ANT in whole plants at a certain day and day 0, respectively

Anthrone and anthraquinone were the primary two ANT metabolites reported by Cajthaml et al. (2002) and Wild et al. (2005). These two metabolites were also observed in ANT-treated tall fescue, but they were undetectable in control plants without ANT. This undoubtedly demonstrated ANT metabolism in the plants, and anthrone and anthraquinone came from inner plant ANT metabolism. The concentrations of anthrone and anthraquinone in roots and shoots of tall fescue during the sampling period are listed in Table 8-1. In general, anthrone concentrations tended to first increase by day 2, dropped at day 4, increased again at day 8, and dropped eventually at day 16 both in roots and shoots. Enhanced anthrone concentrations at the first stage theoretically resulted from ANT degradation in plants, and the last decrease of anthrone would be ascribed to its further metabolism. In contrast, the concentrations of anthraquinone in roots and shoots decreased clearly from 0 day to 16 days, indicating the significant further degradation of anthraquinone derived from ANT metabolism in the plants. Anthraquinone was reported to be degraded into its primary metabolites of hydroxyanthraquinone and phthalic anhydride (Cajthaml et al., 2002; Wild et al., 2005).

Table 8-1 Concentrations of anthrone and anthraquinone in root and shoot of tall fescue /(mg/kg)

Concentrations of anthrone and anthraquinone in plant/(mg/kg)	Time/day				
	0	2	4	8	16
Concentrations of anthrone in root	1.165 a	2.450 c	1.976 b	2.259 bc	1.272 a
Concentrations of anthrone in shoot	1.173 bc	1.480 d	0.857ab	1.222 cd	0.512 a
Concentrations of anthraquinone in root	1.166 c	0.627 bc	0.475 b	0.248 a	0.149 a
Concentrations of anthraquinone in shoot	0.726 c	0.439 b	0.230 a	0.200 a	0.267 ab

Note: Values in the same line followed by the same letter are not significantly different ($p<0.05$).

8.1.2 Distribution of anthracene and its metabolites in subcellular tissues

The measured concentrations (C_{cell}, mg ANT per kg cell wall or organelle on a dry weight basis) of ANT in the subcellular fractions of tall fescue roots and shoots are shown in Figure 8-3. ANT concentrations in the cell walls and organelles decreased significantly during the period from 0 day to 16 days after the ANT-treated seedlings were transplanted to PAH-free solution. This trend was consistent with the patterns of ANT in tall fescue roots and shoots shown in Figure 8-1. ANT concentrations in the cell walls and organelles of roots were 0.012 mg/kg and 0.078 mg/kg at 16 days, which were 96.8% and 93.7% lower than those at 0 day. ANT in shoots underwent a similar pattern: ANT concentrations in the cell walls and organelles of shoots decreased from 0.024 mg/kg and 0.087 mg/kg to 0.019 mg/kg and 0.025 mg/kg between 0 day and 16 days, respectively. The reduced subcellular concentrations of ANT may be due to degradation in shoots and roots, as stated previously. Additionally, ANT concentrations in cell organelles were always much higher than those in cell walls, in both roots and shoots. As shown in Figure 8-3, ANT concentrations in organelles of roots and shoots were 233%~567% and 31%~271% higher than those in the cell walls, respectively, indicating a significant partition of ANT into organelles from the cell solution.

ANT was distributed into three subcellular fractions in roots and shoots: a water-soluble fraction, cell walls, and organelles. The proportions (P, %) of ANT in each subcellular fraction were calculated as follows: $P\ (\%) = M_{cell}/M_p \times 100\%$, where M_{cell} (unit: mg ANT in a certain subcellular fraction of root or shoot) is the total amount of ANT in a certain subcellular fraction of root or shoot, and M_p (unit: mg

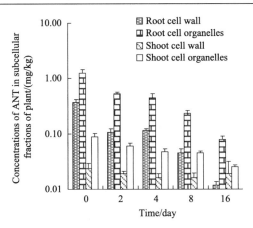

Figure 8-3　Concentrations of anthracene (ANT) in subcellular fractions of roots and shoots of tall fescue as a function of time

ANT in a root or shoot) is the total amount of ANT in root or shoot, respectively. The proportions are plotted against time in Figure 8-4. The percentage of ANT in the cell soluble fraction of roots and shoots decreased dramatically over the sampling period. At 0 day, 33.6% and 88.6% of the ANT was present in the cell soluble fraction of roots and shoots, respectively, but the proportion decreased, and at 16 days no detectable ANT was found in this fraction in either roots or shoots. The ANT decrease in the cell soluble fraction may have resulted from the metabolism and partition into cell solid fractions. This suggestion was supported by the measured proportions of ANT in cell walls and organelles. In contrast to the ANT trend in the cell solution, the percentage of ANT in organelles was enhanced significantly during the sampling period. For example, 43.0% and 7.7% of ANT was present in organelles of roots and shoots, respectively, at 0 day whereas the proportions increased to 78.6% and 42.0% at 16 days. ANT underwent a similar trend in the shoot cell wall: the percentage of ANT in cell walls of shoots increased from 3.7% to 58.0% during the experiment. In contrast, the proportions of ANT (average, 23.6%) in root cell walls changed little from 0 day to 16 days, although ANT concentrations in this fraction decreased extensively. The increased proportions of ANT in cell walls and/or organelles seemed to correspond to the decrease in the cell soluble fraction, suggesting a significant partition of ANT from cell solution into cell solids.

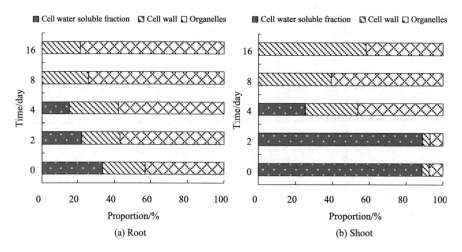

Figure 8-4　Proportions of anthracene (ANT) distributed in the cell water soluble fraction, cell walls, and organelles of roots (a) and shoots (b) as a function of time. The proportion of ANT was calculated from the measured amount of ANT in each fraction to its total amount in plant cells

At 8~16 days after transplantation from the ANT-spiked culture solution to a clean solution (free of ANT), the cell walls and organelles were the dominant storage compartments for ANT in tall fescue roots and shoots, and a negligible portion of ANT was present in plant cell soluble fractions (Figure 8-4). Notably, organelle content was smaller in tall fescue roots and shoots. For example, cell organelle content was only 5.98% in test plant roots; however, the relatively smaller organelle fraction accumulated a similar and/or higher amount of ANT compared with that in cell walls at 8~16 days, probably due to the higher lipid content of organelles.

Plant uptake of PAHs is necessarily accompanied by water from the transpiration stream, and water moves into root systems through apoplastic and symplastic pathways (Gao and Collins, 2009; Gao et al., 2011). Based on the observed ANT in cell walls and organelles and the soluble fraction of tall fescue roots, we postulated that the entrance and distribution of ANT in the inner cells of roots proceeds as follows: ANT in apoplastic and symplastic water passes through cell walls, and portions are captured in the walls, dissolved in cell solution, and partitioned into cell organelles. This proposed process was supported by the decreased percentage of ANT in the cell soluble fraction and the corresponding increase of its percentage in cell solid fractions during the 16-day period after transplantation (Figure 8-4).

The detected high concentrations and proportions of ANT in organelles were derived from ANT partitioning from the cell solution. Hydrophobic organic chemicals with low solubility, such as PAHs, are readily partitioned into a solid phase in a solid-solution system (Li et al., 2005; Kang et al., 2010). The partitioning ability depends on the chemical characteristics, solid composition, and solution conditions. Lipophilic organic chemicals are more readily partitioned into solids, which have higher lipid content (Collins et al., 2006). In this case, the separated soluble cellular components, mainly consisting of the cell solution and largely concentrated in the cell matrix between cells or organelles, can be regarded as an intracellular buffering distribution phase. Because of the hydrophobicity of ANT, these aqueous substances are not easily enriched, and ANT is preferentially located in organelles that have lipophilic components (Gao et al., 2011), resulting in high concentrations in this subcellular fraction. In contrast, lipids in plant cell walls are composed mostly of polysaccharides (90%), with a few structural proteins, lignin, lectin, and mineral elements as well as a small lipid component. The lipid content of plant organelles ranges from 15%~30% (Li, 2006b), enabling the organelles to draw ANT from the cell wall. Thus, the relatively smaller organelle fraction with higher lipid content exhibited higher accumulation ability of ANT than that of cell walls.

However, the ANT concentrations in each cell fraction of tall fescue roots and shoots decreased from 0 day to 16 days, suggesting significant metabolism of ANT in these cell fractions. The measured anthrone and anthraquinone in these cell fractions at 0 day to 16 days supported this notion.

The subcellular distribution of primary ANT metabolites including anthrone and anthraquinone in tall fescue was also investigated. Figure 8-5 shows the anthrone and anthraquinone concentrations (C_{cell-M}, mg anthrone or anthraquinone per kg cell wall or organelle on a dry weight basis) in the cell walls and organelles of plant roots and shoots during the sampling period. Generally, anthrone concentrations in the cell walls and organelles of roots and shoots increased at first but then decreased. The first increase in anthrone concentrations in these subcellular fractions was theoretically ascribed to the ANT metabolism and anthrone partition in the cell solid fractions. The decreased concentrations of anthrone in these portions at the second stage indicated further anthrone degradation. Compared with the concentrations in the cell wall, anthrone concentrations in organelles of roots and shoots were 342%~1153% and 45%~1223% higher, respectively. This result was consistent with the ANT concentrations observed from 0 day to 16 days in these subcellular fractions. Anthraquinone concentrations in cell walls and organelles of roots and shoots generally decreased over the 16 days period, with the exception of shoot cell organelles at 8 days and 16 days. Similar to anthrone, anthraquinone concentrations in organelles of roots and shoots were 55%~497% and 13%~754% higher, respectively, than those in cell walls.

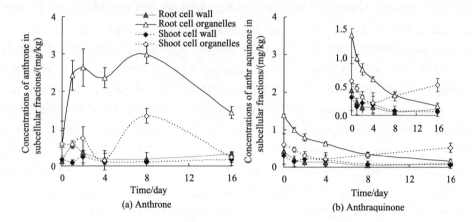

Figure 8-5 Concentrations of anthrone (a) and anthraquinone (b) in subcellular fractions of roots and shoots of tall fescue as a function of time

The proportions (P_M, %) of anthrone and anthraquinone in each subcellular fraction were calculated as follows: P_M (%) = M_{cell-M} / M_{p-M} × 100%, where M_{cell-M} (unit: mg anthrone or anthraquinone in a certain subcellular fraction of root or shoot) is the total amount of anthrone or anthraquinone in a certain subcellular fraction of root or shoot, and M_{p-M} (unit: mg anthrone or anthraquinone in a root or shoot) is the total amount of anthrone or anthraquinone in root or shoot, respectively). The proportions of anthrone and anthraquinone in the cell soluble fraction, cell walls, and organelles of tall fescue roots and shoots versus time are shown in Figure 8-6. Compared with the anthrone and anthraquinone percentages in the cell soluble fraction at 0 day, the proportions in this subcellular fraction were smaller at 16 days, in both roots and shoots. In contrast, the proportions of anthrone in root cell organelles and of anthraquinone in shoot cell organelles were larger at 16 days. Overall, the cell soluble fraction and the organelles were the two major storage locations for anthrone during the 16 days period. At 16 days, the distribution of anthrone was ordered in root cells as cell organelles > cell soluble fraction > cell wall, and in shoot cells as cell soluble fraction ≫ cell organelles ≅ cell walls. In contrast, cell walls and organelles were the two dominant storage compartments of anthraquinone at 16 days, and negligible portions were present in cell soluble fractions. In addition, more anthraquinone was present in cell organelles of shoots (82.5%), whereas cell walls and organelles of roots contained 53.9% and 46.1%, respectively, at 16 days. The percentages of anthrone in cell walls and organelles of roots and shoots were relatively smaller than those of anthraquinone, whereas the percentages of anthrone in the cell soluble fraction of roots and shoots were much larger than those of anthraquinone, possibly due to the higher water solubility of anthrone compared to that of anthraquinone.

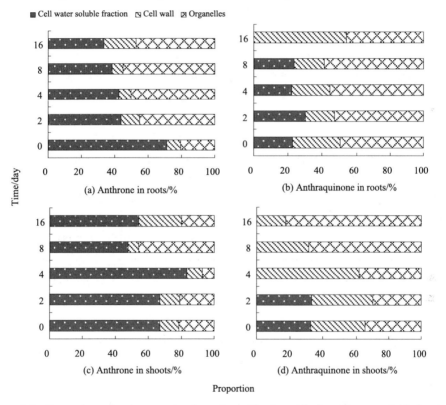

Figure 8-6 Proportions of anthrone and anthraquinone distributed in the cell water soluble fraction, cell walls, and organelles of roots and shoots as a function of time. (a) anthrone in roots, (b) anthraquinone in roots, (c) anthrone in shoots, (d) anthraquinone in shoots. The proportions of anthrone and anthraquinone were calculated from their measured amounts in each fraction to their total amounts in plant cells

The measured primary metabolites of ANT in each intracellular fraction of tall fescue may be derived from the local degradation of ANT in this fraction and transported from other cell parts after metabolism. The significant losses of anthrone and anthraquinone in subcellular fractions suggested their further degradation in the plant, although further metabolites could not be examined due to technical limits till now.

8.1.3 Metabolism mechanism discussion

Metabolism of organic compounds, particularly pesticides, involves a three-phase process (Van Eerd et al., 2003). During phase I metabolism, the initial properties of the parent compound are transformed through oxidation, reduction, or hydrolysis to produce a more water-soluble and usually a less toxic product than the parent. The second phase involves conjugation of a chemical or chemical metabolite into a sugar, amino acid, or glutathione, which increases water solubility and reduces toxicity compared with the parent compound. Generally, phase II metabolites have less or no phytotoxicity. The third phase involves conversion of phase II metabolites into secondary conjugates, which are also nontoxic (Brady et al., 2003). In this investigation, the two primary metabolites of ANT, i.e., anthrone and anthraquinone, were observed in roots and shoots of tall fescue. Because no anthrone or anthraquinone was found in plants without ANT pretreatment, these two metabolites undoubtedly resulted from ANT transformation in the plant. This result was also supported by the direct observation of these metabolites in plants by Wild et al. (2005). Changes in anthrone and anthraquinone concentrations in roots and shoots of tall fescue clearly demonstrated their further transformation inside the plant. This means that after being transformed into anthrone and anthraquinone, ANT was further transformed into secondary metabolites or totally mineralized in the plant.

Additionally, following a rapid increase at 0~4 days, the dissipation ratios of ANT in tall fescue reached a stable state at about 90% after 8 days (Figure 8-2), indicating that about 10% of the ANT could not be readily degraded in the plants. Organic compounds can present in different forms in plants (Li et al., 2001). A portion of the organic compounds may be constructed in plant tissues and form the bound residues that could not be extracted with organic solvent. These unextractable bound residues are difficult to metabolize, possibly resulting in the 10% ANT residue in tall fescue after 8 days.

Metabolism of persistent organic pollutants (POPs) by animals and microorganisms (bacteria and fungi) occurs via enzyme systems (Cajthaml et al., 2002). However, there is very limited information on the enzyme-dominated mechanism of POP metabolism in inner plant bodies (Muratova et al., 2009b). Köller et al. (2000)

reported that applying horseradish peroxidase (HRP) together with defined amounts of hydrogen peroxide removed 90% of the PCB 9 and 55% of the PCB 52 from an aqueous solution after a reaction period of 220 min. Chroma et al. (2002a) demonstrated that Remazol Brilliant Blue R (RBBR) oxidase plays important roles in PCB transformation by plant cells. Moen et al. (1994) first reported phenanthrene degradation by manganese peroxidase (MnP) in *Phanerochaete chrysosporium* as a lipid peroxidation-dependent process. MnP from *Naematoloma frowardii* degraded directly ANT, phenanthrene, pyrene, fluoranthene, and benzo[a]pyrene, leading to partial mineralization (Hofrichter et al., 1998). Cajthaml et al. (2002) reported that *Irpex lacteus* transforms ANT *in vivo* and splits the aromatic rings, and that MnP is the enzyme involved in PAH degradation by whole fungal cultures. Polyphenol oxidase (PPO) and Peroxidase (POD) were reported to play important roles in the metabolism of aromatic chemicals in soil and water environments and had high activities in PAH-treated plants (Mayer, 2006; Karim and Husain, 2010; Saby John et al., 2011; Ling et al., 2012). In the present study, it may be reasonably postulated that ANT degradation in tall fescue is similarly due to enzyme systems, which needs further elucidation in future.

8.2 Enzyme activity in tall fescue contaminated by PAHs

Researchers have postulated that enzymes play major roles in POP degradation in plants (Collins et al., 2006), but, in contrast to bacteria, knowledge of plant enzymes involved in metabolism of POPs is limited, and little information is currently available on enzyme-dominated mechanisms of POP metabolism within plants. Understanding enzyme responses to POP contamination is a key step and of paramount importance in the elucidation of POP metabolic mechanisms in plants.

Polyphenol oxidase (PPO), present in various fungi and plant tissues, is a generic term used for a group of enzymes that catalyze the oxidation of different phenolic compounds (Saby John et al., 2011). Based on observations from fungi and bacteria in microcosm and *in vitro* experiments, PPO was documented to play important roles in the metabolism of aromatic chemicals in soil and water environments (Edwards et al., 1999). However, knowledge is still limited on the PPO response to POP contamination in plants, and few documents are available on PPO activity in subcellular fractions of POP-contaminated plants.

In this section, we investigated PPO activity in tall fescue (*Festuca arundinacea*

Schreb.) contaminated with phenanthrene, a PAH with three fused benzene rings (Ling et al., 2012). Seedlings were incubated in half-strength Hoagland solution until they reach approximate 15 cm height and relatively mature roots had developed. Then seedlings were transplanted into half-strength Hoagland solution with phenanthrene (the concentrations of phenanthrene in solution after 144 h were 0.0012~0.384 mg/L), and hydroponically cultured for 144 h. Then seedlings were sampled, washed with 1 L Milli-Q water, blotted with tissue paper, separated for roots and shoots, and prepared for PAH and enzyme analysis. The subcellular fractionation of plant cells followed Weigel and Jager (1980), Kang et al (2010), and Gao et al (2011). PPO analysis referred to Hao et al. (2004) and Muratova et al. (2009b). Results of this work provide some fundamental information on POP metabolic mechanisms in plants, and are valuable for plant contamination risk assessment.

8.2.1 Enzyme activity in tall fescue

Before measurement of enzyme activity in tall fescue, the growth and development of plants under contamination stress were evaluated to assess the phytotoxicity of phenanthrene. It was found that tall fescue grown in culture solution with phenanthrene at test concentrations did not show any outward signs of stress or phytotoxicity. There were no significant differences between root and shoot biomass in phenanthrene-spiked solutions and those in phenanthrene-free controls. The phytotoxicity of PAHs is mainly determined by their lipophilicity, water solubility, and bioavailability, and also closely related to the plant species examined (Thygesen and Trapp, 2002; Alkio et al., 2005). Plants exhibit responses to toxic compounds only after being exposed to a "threshold concentration", which varies for different compounds and plant species (Muratova et al., 2009b). Here, no visible phytotoxic effects were found, indicating that tall fescue was less sensitive to phenanthrene contamination at the test levels used.

PPO activity in tall fescue roots and shoots (E_{plant}) after phenanthrene contamination is given in Figure 8-7. After 144 h of cultivation in solutions with 0~0.38 mg/L phenanthrene, the PPO activity in roots and shoots ranged from 4957 U/mg to 6318 U/mg root and 1187 U/mg to 2295 U/mg shoot, respectively. On the whole, phenanthrene at concentrations below 0.11 mg/L did not induce significant changes of PPO activity in the plants. In contrast, phenanthrene contamination above 0.23 mg/L significantly stimulated PPO activity in roots and shoots. Compared with the activity of PPO in shoots, root PPO activity was 153%~359% higher for the same treatment.

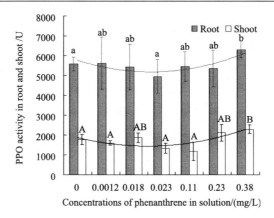

Figure 8-7 Polyphenol oxidase (PPO) activity in roots and shoots of tall fescue as a function of phenanthrene concentration in solution. Values above columns followed by the same letter are not significantly different ($p < 0.05$)

This investigation showed that phenanthrene contamination at a relatively high level (>0.23 mg/L) in solution stimulated PPO activity in tall fescue roots and shoots (Figure 8-7), as reported previously (Criquet et al., 2000). This could be related to the function of enzymes in plants. PPO, as well as other enzymes, has an intracellular function as part of plant defense mechanisms and/or a direct effect on the degradation of aromatics within the plant. When more xenobiotic organic contaminants enter a plant, higher enzyme activity is required to protect the plant from toxicity. However, this is not the case when contaminant levels are high enough to result in phytotoxicity (even death), and hence enzyme activity is sharply reduced. In our previous study, we observed that PPO and peroxidase activities in ryegrass first increased and then decreased with increasing concentrations of phenanthrene and pyrene in solution (Lu et al., 2008), which supports the above hypothesis.

8.2.2 Enzyme activity in subcellular fractions of tall fescue

Roots and shoots of tall fescue were fractionated into three subcellular fractions: cell solution (cell water and water-soluble fraction), cell wall, and organelles. The amount of PPO activity in each cell fraction (E_{cell}) is given in Table 8-2. PPO activity in cell solution, cell wall, and organelles of roots under the impact of 0~0.38 mg/L phenanthrene was 4081~5257 U/mg, 646~950 U/mg and 100~153 U/mg root and activity in subcellular fractions of shoots was 961~1882 U/mg, 190~366 U/mg and 8.1~46.5 U/mg shoot, respectively. The amount of PPO activity in the three cell fractions per mg root or shoot was, in descending order: cell solution > cell wall > cell organelles. Moreover, PPO activity in root cell solution, cell wall, and organelles was 179%~370%, 155%~281% and 53%~1217% higher than in the corresponding shoot cell fractions. On a whole, PPO activity in root and shoot cell walls generally increased with increasing solution concentration of phenanthrene from 0 mg/L to 0.38 mg/L. The weak contamination stress with phenanthrene concentrations of lower than 0.23 mg/L did not result in significant changes in test enzyme activity in root and shoot cell solution and organelles, while 0.38 mg/L phenanthrene might enhance PPO activity in these cell fractions.

Table 8-2 The amount of polyphenol oxidase (PPO) activity in subcellular fractions (E_{cell}) of tall fescue contaminated with phenanthrene

	Concentrations of phenanthrene in solution/(mg/L)	PPO activity in subcellular fractions/(U/mg)		
		Cell solution	Cell wall	Cell organelle
Root	0	4821.2±1203.2	657.7±93.8	100.0±35.0
	1.2×10^{-3}	4851.0±282.4	645.5±40.0	131.4±33.3
	18.1×10^{-3}	4593.4±1007.6	684.7±30.8	152.7±116.0
	23.1×10^{-3}	4080.5±723.5	773.2±103.6	103.6±29.5
	109.3×10^{-3}	4513.5±567.0	830.7±178.9	106.7±16.1
	234.0×10^{-3}	4345.7±799.4	851.5±101.3	59.2±27.9
	384.3×10^{-3}	5256.8±185.7	950.4±203.7	110.4±27.2
Shoot	0	1495.2±122.2	237.2±27.4	30.0±10.1
	1.2×10^{-3}	1372.8±74.8	202.6±39.0	23.9±4.3
	18.1×10^{-3}	1353.3±199.0	190.0±36.3	21.6±9.7
	23.1×10^{-3}	1040.7±138.5	267.3±10.3	16.3±8.4
	109.3×10^{-3}	961.2±427.2	218.0±27.9	8.1±3.9
	234.0×10^{-3}	1744.4±319.8	333.5±72.9	38.6±21.5
	384.3×10^{-3}	1882.2±178.4	365.8±125.9	46.5±5.0

The proportion (P_{cell}) of PPO activity in subcellular fractions of tall fescue roots and shoots was calculated according to the equation

$$P_{cell}(\%) = E_{cell} / E_{plant} \times 100$$

where E_{cell} is the amount of PPO activity in a certain cell fraction per unit (mg) root or shoot, and E_{plant} is the total amount of PPO activity per unit (mg) root or shoot. A higher P_{cell} value means a larger portion of PPO present in this root or shoot cell fraction. P_{cell} values are displayed in Figure 8-8. As shown, cell solution was the dominant fraction for PPO in roots and shoots, averaging 84.0% and 82.8% of PPO activity present in cell solution, respectively. In contrast, 11.5%~16.3% (average 14.1%) and 12.1%~20.3% (average 15.7%) of PPO in the root and shoot was present in cell walls, respectively. While a negligible portion, less than 2.6% in roots and 1.9% in shoots, of PPO was observed in the cell organelle fraction. Overall, the distribution of PPO in each cell fraction of tall fescue roots and shoots, irrespective of solution contamination level, was in descending order, cell solution≫cell wall>cell organelles.

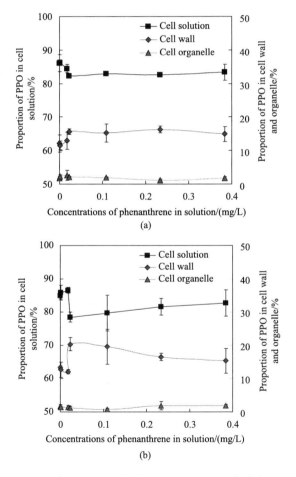

Figure 8-8 The proportion of PPO in subcellular fractions (P_{cell}) of tall fescue roots (a) and shoots (b) as a function of phenanthrene concentration in solution

The contents of the cell fractions differed greatly in tall fescue, and the cell wall and organelle contents were smaller than that of the cell solution. For example, the content of cell walls and organelles was only 8.92% and 5.98% in test plant roots, while the content of cell solution was 85.1%. E_{cell} clarified the amount of PPO activity in the cell fraction per unit (mg) root, but did not provide information on the density of PPO activity per unit (mg) cell fraction. To this end, the cell fraction content normalized PPO activity (E_{cell-n}) in subcellular fractions of tall fescue was introduced, and calculated as

$$E_{cell-n} = E_{cell} / f_{cell}$$

where f_{cell} (%) is the content of a cell fraction in tall fescue root or shoot. A higher E_{cell-n} value means that one unit (mg) of a certain cell fraction has higher PPO activity (Figure 8-9). Clearly, although the content of cell walls in tall fescue was much smaller than that of the cell solution, the E_{cell-n} values of PPO in cell walls were much larger than those in the cell solution, indicating that one unit of cell wall had higher PPO activity than the cell solution. In addition, the E_{cell-n} values in cell walls overall increased with the increase of phenanthrene concentration in the solution. This was also the case for PPO in the cell solution and organelles at phenanthrene concentrations above 0.23 mg/L. This confirms that high phenanthrene contamination stimulated PPO activity in subcellular fractions of the plant. The E_{cell-n} values in all three cell fractions of tall fescue roots and shoots were in descending order: cell wall > cell solution > cell organelles. This means that, per unit, cell walls contained more PPO than the other two cell fractions.

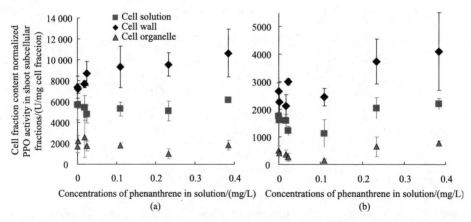

Figure 8-9 Cell fraction content-normalized polyphenol oxidase (PPO) activity in subcellular fractions of tall fescue roots (a) and shoots (b) as a function of phenanthrene concentration in solution

One notes that, to our knowledge, this is the first study to show enzyme activity in plant subcellular fractions after contamination stress from PAHs. Tall fescue roots and shoots were fractionated into three subcellular fractions, cell solution, cell wall, and organelle, as in previous studies (Weigel and Jager, 1980; Gao et al., 2011). The recovery rates (R, %) of PPO in intracellular fractions were measured, and R calculated as

$$R(\%) = (E_{\text{cell-solution}} + E_{\text{cell-wall}} + E_{\text{cell-organelle}}) / E_{\text{plant}} \times 100$$

where $E_{\text{cell-solution}}$, $E_{\text{cell-wall}}$ and $E_{\text{cell-organelle}}$ are the amount of PPO activity in the cell solution, cell wall, and organelles per unit (mg) root or shoot, respectively. The measured R values were between 96%~105%, indicating that the fractionation method used was reliable and caused negligible damage to PPO activity.

It has been speculated that several nonspecific enzymes dominate the metabolism of POPs in plants, and that these enzymes may be within the cell vacuole, plasmalemma, and cell walls (Wild et al., 2005). However, little information is available on enzyme activity in intracellular fractions of plants under PAH contamination. The present work revealed that the cell solution was the dominant storage domain of PPO and contributed 84.0% (roots) and 82.8% (shoots) of PPO activity in phenanthrene-contaminated tall fescue. However, the cell wall had the highest density of PPO activity (per unit subcellular fraction). Since PPO plays important roles in the degradation of phenanthrene (Edwards et al., 1999; Gong et al., 2010), it could reasonably be postulated that the cell solution and cell wall were the dominant metabolic domains for phenanthrene in roots and shoots of tall fescue, given the high measured values of E_{cell}, P_{cell}, and $E_{\text{cell-n}}$ in these two cell fractions.

8.3 Inhibitor reduces enzyme activity and enhances PAH accumulation in tall fescue

Inhibitors are commonly utilized in agricultural production to control enzyme activities and the metabolism of organic components, such as herbicides, by plants. Sterling and Balke (1990) reported reduced effects of monooxygenase inhibitors (1-aminobenzotriazole, tetcyclasis, piperonyl butoxide and cinnamic acid) on the oxidative metabolism of bentazon in rice and in soybean cell cultures. This inhibition has also been reported for the herbicides fenoxaprop-ethyl and diclofop methyl in wheat or barley by tetcyclacis and tridiphane (Romano et al., 1993). Gronwald and Connelly (1991) reported that the cytochrome P450 inhibitor phenylhydrazine significantly diminished bentazon metabolism; moreover, other inhibitors–such as 3(2,4-dichlorophenoxy)-1-propyne and aminobenzotriazole also reduced bentazon metabolism, albeit to a lesser extent. A significant decrease in the activities of arginine decarboxylase and ornithine decarboxylase occurred when DL-α-difluoromethylornithine (DFMO) and DL-α-difluoromethylarginine (DFMA) were added to a maize (*Zea mays* L.) callus culture medium, and resulted in irreversible inhibition of putrescine synthesis (Torné et al., 1994). However, to our knowledge, most previous studies of the effects of enzyme inhibitors on plant metabolism focused primarily on herbicide applications in agricultural production, and little is known about the effects of inhibitors on plant metabolism of absorbed POPs such as PAHs.

Ascorbic acid (AA) is a naturally occurring, water-soluble compound with desirable characteristics as an enzyme inhibitor. It is the most abundant antioxidant in plants, and is used in agriculture to enhance plant stress-resistance (Smirnoff, 2000). A recent *in vitro* study reported that AA inhibits the activity of PPO in *Mangifera indica* L. (Saby John et al., 2011). However, few studies have investigated the effects of AA on enzyme activities and the metabolism of PAHs by plants. To this end, the objective of this study was to evaluate the influence of the commonly used inhibitor, AA, on plant enzyme activities and PAH uptake (Gao et al., 2012).

Naphthalene (NAP), PHE, and ANT, as representative 2-ring and 3-ring PAHs, were the PAHs used. Polyphenol oxidase (PPO) and Peroxidase (POD) play important roles in the metabolism of aromatic compounds in soil and water (Baborová et al., 2006; Karim and Husain, 2010). *In vitro* degradation of PAHs by POD and PPO was conducted by mixing 1.0 mL POD or PPO stock solution (PBS and Tris-HCl buffers) with 9.0 mL PAH solution. The final concentrations of NAP, PHE, and ANT were 10 mg/L, 1.0 mg/L and 0.04 mg/L, respectively (Gao et al., 2012). Control treatments were conducted using the same amounts of PBS or Tris-HCl buffer and PAHs but in the absence of POD or PPO enzymes. Plant uptake of PAH from water with/without AA was processed as follows. Seedlings of tall fescue (*Festuca arundinacea* Schreb.) were incubated in half-strength Hoagland solution until they reach approximate 15 cm height and relatively mature roots had developed. Then seedlings were transplanted into Half-strength Hoagland solution containing 1.0 mg/L PHE and/or 2.0 mg/L AA. The control treatments were prepared in the same manner but without AA. After plant uptake for 1 days, 2 days, 4 days, 8 days and 16 days, the plants were sampled and prepared for analyses of PHE concentration and enzyme activities (Gao et al., 2012). Findings of this investigation suggest that the common use of enzyme inhibitors in agricultural production may promote the accumulation of organic contaminants in plants, hence increasing risk in terms of food safety and quality.

8.3.1 *In vitro* degradation of PAHs in solution with enzymes

The *in vitro* degradations of NAP, PHE, and ANT in the presence of POD and PPO enzymes are shown in Figure 8-10, along with the enzyme-free controls for comparison. The degradation fraction (D_{PAH}, %) of PAHs was estimated

$$D_{PAH} = (C_{PAH\text{-}o} - C_{PAH\text{-}i}) / C_{PAH\text{-}o} \times 100\%$$

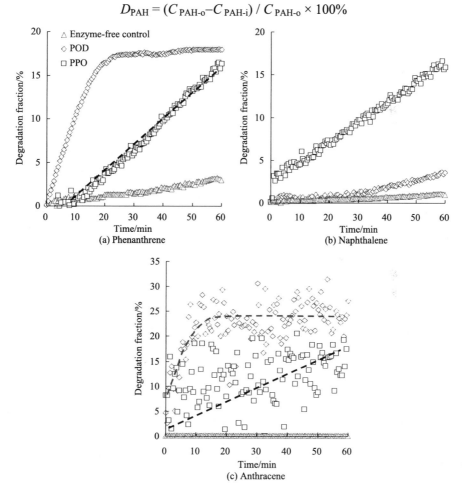

Figure 8-10 In vitro degradation fraction (%) of added (a) phenanthrene, (b) naphthalene, and (c) anthracene by POD and PPO over time. The degradation fraction (D_{PAH}, %) of PAHs was estimated as: $D_{PAH} = (C_{PAH\text{-}o} - C_{PAH\text{-}i}) / C_{PAH\text{-}o} \times 100\%$, where $C_{PAH\text{-}i}$ and $C_{PAH\text{-}o}$ represent the PAH concentrations in solution at time i and time 0, respectively

where C_{PAH-i} and C_{PAH-o} are the PAH concentrations in solution at time i and time 0. This *in vitro* experiment clearly demonstrated that the presence of POD and PPO in the solution facilitated the decomposition of PAHs with varying degradation rates. In the control treatments, the concentration of the test PAHs decreased by less than 3.1% during the experimental period, possibly due to abiotic dissipation processes such as sorption to glass walls and/or volatilization. PAH degradation in the presence of POD and PPO was significantly greater than that in the enzyme-free controls. Approximately 17.9% and 16.5% of PHE, 3.52% and 16.5% of NAP, and 34.7% and 16.3% of ANT were decomposed within 60 min, indicating that POD and PPO enzymes degrade PAHs rapidly under these conditions.

POD and PPO exerted different effects in terms of degradation of PAHs *in vitro*. Rapid decomposition of PHE and ANT by POD occurred. The degradation fraction increased sharply to 17.2% and 26.3% within the first 20 min, and thereafter approached a steady state from 20~60 min. No apparent difference in NAP degradation in the first 30 min was observed for POD *vs.* the POD-free control, but NAP degradation increased slightly to 3.53% from 30~60 min. Clearly, POD degraded PHE and ANT more rapidly than NAP. In contrast, degradation of the three PAHs by PPO exhibited a linear increase; up to ~18% was degraded after 60 min. No apparent differences in degradation rates or magnitudes were found among the three PAHs, indicating that PPO is not selective in terms of its degradation of the studied PAHs.

Our data, together with several recent reports, indicate that PPO and POD facilitate the decomposition of PAHs and other aromatic hydrocarbons (Baborová et al., 2006; Karim and Husain, 2010; Saby John et al., 2011). Moen et al. (1994) first reported metabolism of PHE by manganese peroxidase (MnP) in *Phanerochaete chrysosporium* in a lipid peroxidation-dependent process (Moen and Hammel, 1994.). MnP from *Nematoloma frowardii* degraded ANT, PHE, pyrene, fluoranthene, and benzo[a]pyrene, leading to the partial mineralization of these persistent compounds (Sarkar et al., 1997; Hofrichter et al., 1998). The MnP of *Irpex lacteus* efficiently decomposed 3-ring and 4-ring PAHs including PHE and fluoranthene (Bogan and Lamar, 1995). *Irpex lacteus* in whole fungal cultures could transform ANT and open its aromatic rings *in vivo*, for which MnP is the essential enzyme (Cajthaml et al., 2002).

In our study, POD preferentially decomposed ANT and PHE rather than NAP *in vitro*. This is in contrast to the dogma that larger fused-ring PAHs are more resistant to degradation. PPO decomposed PAHs, but was not specific for a particular compound.

POD and PPO are large proteins (33~200 kDa for PPO, and 35~105 kDa for POD) (Sherman et al., 1991; Saby John et al., 2011). The catalytic degradation of organic chemicals by enzymes depends on the enzyme structure and complex formation. Henriksen et al. (1999) reported that horseradish peroxidase forms peroxidase and ferulic acid binary complexes and peroxidase, cyanide and ferulic acid ternary complexes, which contain the flexible aromatic donor binding sites. The complex component ferulic acid is a naturally occurring phenolic compound commonly found in plant cell walls. The flexible binding sites may interact with organic chemicals such as PAHs and initiate decomposition. The more hydrophobic nature of PHE and ANT and their relatively large size might lead to stronger binding sites than that of NAP; this is presumed to be responsible for the preferential decomposition of PHE and ANT. Exotic organic contaminants (e.g., PAHs) in plants also impact enzyme activity (Lu et al., 2008). The relatively high concentration of NAP (10 mg/L) used might have reduced POD activity, leading to negligible NAP degradation by POD at 0~30 min and only slight degradation at 30~60 min. Overall, the mechanism of decomposition, and particularly the selectivity of the POD enzyme, remains unclear, and is worthy of further study.

8.3.2 Effects of inhibitor on enzyme activities in plants

POD and PPO activities in tall fescue roots were measured in the presence and absence of AA for 16 days (Figure 8-11). Activities decreased over time irrespective of the presence of AA. For instance, POD and PPO activity in the roots of the control treatments was 1.11 katal/mg and 0.59 katal/mg root, respectively, at day 1, but decreased to 0.58 katal/mg and 0.47 katal/mg root, respectively, at day 16. Similar decreases in enzyme activity over time also occurred in the presence of AA. However, the addition of AA led to marked reductions in enzyme activities in plant roots. During the study period (16 days), POD and PPO activities in roots in the presence of AA were approximately 26.5%~83.2% and 29.1%~89.1% lower, respectively, than those of the controls. Furthermore, the presence of AA resulted in rapid decreases in enzyme activity during the first 8 days, after which a relatively steady state was maintained in 8~16 days. On day 16, the POD and PPO activities in the presence of AA were 0.10 katal/mg and 0.05 katal/mg root, respectively. Therefore, 2.0 mg/L AA inhibited the POD and PPO activities in plant roots.

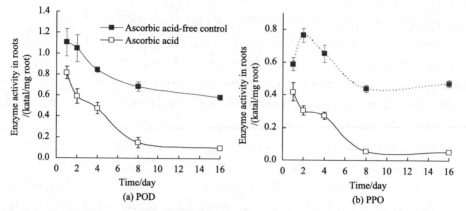

Figure 8-11 (a) POD and (b) PPO activities in the roots of tall fescue over time. The uptake time was defined as the time frame from the immersion of plant roots in solution containing phenanthrene to the removal of plant for extraction. Error bars represent standard deviation (SD). The initial phenanthrene concentration was 1.0 mg/L

The reductions in enzyme activity were estimated by calculation of E_{enzyme} (%)

$$E_{enzyme} = (U_c - U_{AA}) / U_c \times 100\%$$

where U_{AA} and U_c are the enzyme activities in plant roots grown in the presence and absence, respectively, of AA. A greater E_{enzyme} value indicates greater inhibition of enzyme activity. Figure 8-12 shows the E_{enzyme} values of the test enzymes in roots as a function of time. The E_{enzyme} values of both POD and PPO, increased dramatically with time, indicating that AA inhibited both enzymes. AA generally inhibited PPO to a greater extent than POD during the study period.

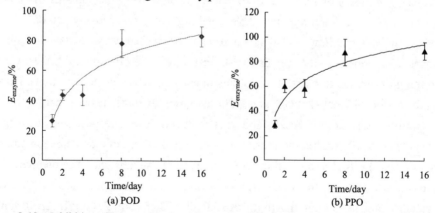

Figure 8-12 Inhibition efficiency (E_{enzyme}, %) of ascorbic acid on (a) POD and (b) PPO activities in plant roots. E_{enzyme} was calculated as: $E_{enzyme} = (U_c - U_{AA}) / U_c \times 100\%$, where U_{AA} and U_c represent the enzyme activities in roots grown in the presence and absence of AA, respectively. Error bars represent standard deviations (SD)

8.3.3 Effects of inhibitor on the enhanced accumulation of PAH in plants

The impact of AA on root PHE uptake (as a representative PAH) was evaluated using the results of the hydroponic experiments. The PHE concentration decreased over the exposure period (Figure 8-13(a)), which may be due to either PHE degradation or metabolism by the plant. Addition of AA enhanced PHE accumulation in the roots of tall fescue. The corresponding PHE concentrations in AA-treated plants were 24.3~44.5 mg/kg, being 27%~313% greater than those (5.88~35.0 mg/kg) of the control. In addition, the PHE concentration in AA-treated plant roots decreased rapidly during the first 8 days, and thereafter approached a steady state at day 8~16. This is consistent with the AA-mediated inhibition of enzyme activity over time (Figure 8-11).

Figure 8-13(b) shows the accumulated mass (M_{PHE}, μg) of PHE in tall fescue roots as a function of uptake time. M_{PHE} was calculated by multiplication of the concentration of PHE in roots (C_{PHE}, μg/g) and plant root mass (M_{root}, g)

$$M_{PHE} = C_{PHE} \times M_{root}$$

Similar to PHE uptake in roots, PHE accumulation decreased over time. Plant growth results in an increase in biomass that may dilute PHE concentration, particularly during rapid growth. PHE mass is a better descriptor of solute uptake because it excludes the effect of dilution. The presence of AA significantly enhanced PHE accumulation by tall fescue roots (Figure 8-13(b)). As evidenced by the reduced PHE uptake in its absence, the presence of AA inhibits plant enzyme activity, and so enhances PHE accumulation in tall fescue roots.

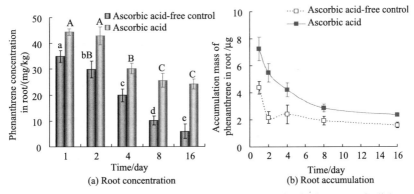

Figure 8-13 (a) Phenanthrene concentration and (b) accumulation in the roots of tall fescue as a function of uptake time. The initial phenanthene concentration was 1.0 mg/L. Error bars represent standard deviations (SD). Values above columns followed by the same letter are not significantly different ($p > 0.05$)

The dissipation of organic compounds in plants is due to the metabolic processes therein (Hannink et al., 2002; Nepovim et al., 2004). The presence of AA inhibits such metabolic processes, leading to enhanced PAH accumulation. We quantified this effect by means of inhibition efficiency (E_{PHE}, %)

$$E_{PHE} (\%) = (C_{PHE-AA} - C_{PHE-c}) / C_{PHE-c} \times 100\%$$

where C_{PHE-c} and C_{PHE-AA} represent the PHE concentrations in plant roots grown without and with AA, respectively. A higher E_{PHE} value indicates greater inhibition of PHE metabolism. The E_{PHE} values increased from 27% to 313% from day 1 to day 16 (Figure 8-14), indicating that the presence of AA persistently inhibited enzyme activities and hence PHE metabolism in plants. Such inhibitory effects became more significant with increasing exposure time.

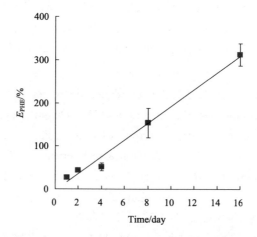

Figure 8-14 Inhibition efficiency (E_{PHE}, %) of ascorbic acid on phenanthrene degradation in plant roots as a function of time. E_{PHE} was calculated as: $E_{PHE} (\%) = (C_{PHE-AA} - C_{PHE-c}) / C_{PHE-c} \times 100\%$, where C_{PHE-c} and C_{PHE-AA} represent the phenanthrene concentrations in plant roots grown without and with AA, respectively. Error bars represent standard deviations (SD)

Enzymes control the decomposition and metabolism of organic chemicals in plants (Gronwal and Connelly, 1991; Romano et al., 1993). AA uptake inhibited enzyme activity and hence diminished PAH metabolism by plants. As a result, enhanced PAH accumulation is expected. Plotting of E_{PHE} values against the E_{enzyme} values of POD and PPO (Figure 8-15) showed a significant and positive correlation between E_{PHE} and E_{enzyme}, which further confirms that AA inhibits POD and PPO activities, reducing PHE degradation. This results in enhanced PHE accumulation in tall fescue roots and higher risk of contamination of the ecosystem. To further confirm this, tall fescue roots were immersed in water at 100°C for 5 min to inhibit enzyme activity. The hot water-treated plants and untreated tall fescue were then exposed to various PHE concentrations. The PHE concentration in roots was measured at 12 h, 24 h and 48 h (Figure 8-16). The PHE concentration in hot water-treated roots (enzyme activity inhibited) was significantly higher than that in untreated roots, confirming that inhibition of enzyme activity results in enhanced PHE accumulation in roots.

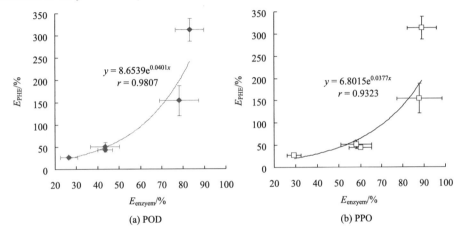

Figure 8-15 Relationship between E_{PHE} and E_{enzyme} for (a) POD and (b) PPO. E_{PHE} and E_{enzyme} are the inhibition efficiencies of ascorbic acid on phenanthrene degradation and enzyme activities in plant roots, respectively. Error bars represent standard deviations (SD)

Our data suggest that the use of external chemicals (i.e., AA) to regulate enzyme activity in plants influences PAH accumulation. This suggests that the application of enzyme inhibitors enhances the PAH-contamination risk. Many commonly utilized inhibitors may influence enzyme activity and hence impact the metabolism of organic chemicals by plants. In this study, AA was shown to inhibit PPO and POD activities, and consequently increase PHE accumulation, in plants. In general, PHE accumulation depends on plant uptake capacity, volatilization, dilution due to growth, and metabolism. PAH uptake by plants is a series of passive partition processes (Chiou et al., 2001), and there was no apparent difference in biomass irrespective of the presence of AA. The volatilization of PHE from plants makes a minor contribution (if any) due to its relatively low K_H (Gao and Ling, 2006). Thus, the difference in PHE accumulation in roots in the presence and absence of AA (Figure 8-13) was likely due to altered enzymatic PHE degradation. The positive relationship between E_{PHE} and

E_{enzyme} (Figure 8-15) indicates that AA inhibits enzymatic PHE degradation, leading to greater residual PHE in plants.

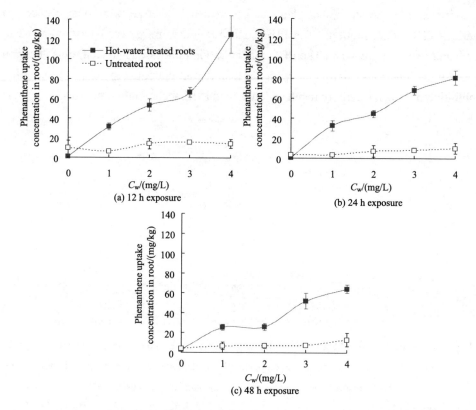

Figure 8-16 Phenanthrene uptake (mg/kg) in tall fescue roots treated by immersion in 100°C water for 5 min, compared to the control. Plant roots were exposed to phenanthrene at initial concentrations of 0~4 mg/L in aqueous solution and sampled after (a) 12 h, (b) 24 h and (c) 48 h. Error bars represent standard deviations (SD)

Chapter 9　Arbuscular mycorrhizal fungi influence PAH uptake by plants

Plant-arbuscular mycorrhizal fungi (AMF) symbioses are ubiquitous in the environment. Arbuscular mycorrhizae (AM) have positive effects on plant establishment and survival in contaminated soils by increasing nutrient uptake, improving drought tolerance, and potentially protecting roots from plant pathogens (Khan et al., 2000; Hart and Trevors, 2005; Hildebrandt et al., 2007). The impacts of AMF inoculation on the uptake of organic pollutants by plants have recently been reported. For instance, Huang et al. (2007) and Wu et al. (2008) found that AMF colonization led to increased accumulation of DDT and atrazine in roots but a decrease in shoots of maize (*Zea mays* L.) and alfalfa (*Medicago sativa* L.). These findings highlight the importance of AMF inoculation to plant uptake of organic contaminants. However, the mechanisms involved in mycorrhizal root uptake of organic chemicals from soil environments are still not fully understood.

AMF colonization results in the formation of an abundance of extraradical hyphae. These hyphae can be 5~50 m in length per gram of soil, which is several orders of magnitude longer than plant roots (Khan et al., 2000). However, very little information is available on the uptake of organic contaminants in soils or soil fine pores by AMF hyphae and their translocation to plant roots. The contributions of AMF hyphae to root uptake of organic chemicals is still under investigation.

In this chapter, the impacts of AMF on plant uptake and accumulation of PAHs in soils were primarily investigated. The function of AMF hyphae in the PAH uptake by plants was evaluated using three-compartment systems. Results of this work provide insight into the behaviors and mechanisms of mycorrhizal root uptake of PAHs in soils and can aid in the assessment of PAH-related risks at contaminated sites.

9.1　PAH uptake by arbuscular mycorrhizal plants

Information about PAH distribution in arbuscular mycorrhizal plants is essential to guide the screening of plants used in arbuscular mycorrhizal phytoremediation (AMPR) and to ensure the security of agricultural products grown on contaminated sites. However, little experimental data are hitherto available on the impacts of AMF on plant uptake of PAHs in soil.

Here, we seek to investigate the impacts of two AMFs, *Glomus mosseae* (AMF1; BGC GD01A) and *G. etunicatum* (AMF2; BGC HUN02C), on the uptake and accumulation of phenanthrene and pyrene as representative PAHs by alfalfa (*Medicago sativa* L.) using a greenhouse study (Gao et al., 2011c). Soil (Typic Paleudalfs) samples were collected from the A (0~20 cm) horizon in Nanjing, China, with a pH of 6.02, 14.3 g/kg soil organic carbon content, 24.7% clay, 61.9% silt and 13.4% sand. The plant growth medium was a 3:1 (w/w) mixture of soil and sand (both sieved through a 2-mm mesh). The soil mix (henceforth referred to as soil) was sterilized by γ-radiation (10 kGy, 10 MeV γ-rays) to inactivate the native AM fungi. Then soils were PAH-spiked, aged, sieved and potted for plant vegetation. The final concentrations of phenanthrene and/or pyrene in soils were 103 mg/kg and 74 mg/kg (on a dry weight basis), respectively. Mycorrhizal pots were inoculated with AMF1 and/or AMF2. The non-mycorrhizal controls received an equivalent amount of γ-radiation-sterilized inoculum to provide similar conditions, except for the absence of the active mycorrhizal fungus. Pre-germinated seeds of alfalfa were sown in each pot, and plants were incubated. Soils and plants were destructively sampled after 30 days, 45 days, 60 days and 70 days, and prepared for PAH analysis.

9.1.1 Arbuscular mycorrhizal colonization of root exposed to PAHs in soil

Joner et al. (2001) reported that colonization by *Glomus mosseae* of clover and ryegrass in soils spiked with anthracene, chrysene, and dibenz[a,h]anthracene was 20%~40% and 0.5%~5.0%, respectively. Here, colonization by the two tested AMF species, *G. mosseae* (AMF1) and/or *G. etunicatum* (AMF2), of alfalfa in soils contaminated with phenanthrene and/or pyrene ranged from 26.4% to 76.5% after 45~70 days (Table 9-1). Since the soils in our greenhouse experiments were sterilized, the colonization of roots by soil indigenous fungi was undetectable. Compared with phenanthrene-spiked soil, soil co-contamination by phenanthrene and pyrene generally reduced mycorrhizal colonization. For instance, the respective colonization of AMF1 on alfalfa in phenanthrene-spiked soils was 56.1%, 55.7% and 60.5% after 45 days, 60 days and 70 days. By contrast, the corresponding colonization in soils co-contaminated with phenanthrene and pyrene was 13.2%~52.9% lower. This indicates enhanced toxicity to mycorrhizae with the addition of the co-contaminant in the soil environment.

Table 9-1 Arbuscular mycorrhizal colonization (%) of alfalfa (*Medicago sativa* L.) exposed to PAH(s) in soil

AMF type	PAHs in soils	Time/day		
		45	60	70
AMF1	Phenanthrene	56.1 (6.73)a	55.7 (11.7)ab	60.5 (4.95)b
AMF2	Phenanthrene	—	—	76.5 (15.5)a
AMF1+AMF2	Phenanthrene	47.5 (8.09)a	63.3 (9.07)a	70.5 (7.78)a
AMF1	pyrene	—	—	66.0 (9.90)ab
AMF1+AMF2	pyrene	—	—	54.0 (14.1)bc
AMF1	Phenanthrene+pyrene	26.4 (7.56)b	48.3 (17.8)ab	52.5 (2.12)c
AMF2	Phenanthrene+pyrene	—	—	61.0 (5.66)ab
AMF1+AMF2	Phenanthrene+pyrene	30.3 (9.99)b	52.3 (4.16)b	41.0 (14.3)d

Note: Data in brackets are standard deviations (SD). AMF means arbuscular mycorrhizal fungi; "—" means not measured. Values in the same columns followed by the same letter are not significantly different ($p<0.05$).

9.1.2 PAH uptake by arbuscular mycorrhizal plants

As shown in Table 9-2 and Figure 9-1, alfalfa clearly took up and accumulated PAHs from spiked soils. Root concentrations of phenanthrene and pyrene were always significantly higher than in shoots, irrespective of inoculation with AMF1, AMF2, or both, indicating restricted translocation of the test PAHs into the plant body. Arbuscular mycorrhizal inoculation significantly impacted alfalfa uptake of the test PAHs from soils. As shown in Figure 9-1, inoculation with *G. mosseae* and/or *G. etunicatum* clearly caused a decrease in PAH concentrations in shoots. The respective concentrations of phenanthrene in shoot from soils inoculated with AMF1 and both AMFs after 45 days, 60 days and 70 days were 18.3%~31.6% and 33.1%~43.1% lower than those in non-mycorrhizal control treatments. In contrast to shoots, AMF-inoculated roots consistently accumulated more phenanthrene than non-mycorrhizal roots. Root concentrations of phenanthrene in plants grown for 70 days in phenanthrene-spiked soils with AMF inoculation were ~40.2% higher than those of non-mycorrhizal controls (Table 9-2). Recently, Wu et al. (2008) investigated the uptake of DDT by alfalfa (*Medicago sativa* L.) after AMF inoculation, and found that mycorrhizal colonization led to increased accumulation of DDT in roots but a decrease in shoots (Wu et al., 2008). The present results agree with these findings, and highlight the importance of AMF inoculation to plant uptake of organic contaminants.

Table 9-2 Concentrations of phenanthrene and pyrene in plant roots and shoots for various treatments after 70 days

AMF	PAHs in soils	Plant concentrations of phenanthrene/(mg/kg)		Plant concentrations of pyrene /(mg/kg)	
		Shoot	Root	Shoot	Root
No AMF	Phenanthrene	0.73 (0.05)c	1.27 (0.23)c	—	—
AMF1	Phenanthrene	0.61 (0.14)cd	1.78 (0.73)bc	—	—
AMF2	Phenanthrene	0.59 (0.03)d	1.33 (0.20)c	—	—
AMF1+AMF2	Phenanthrene	0.59 (0.08)d	1.63 (0.04)b	—	—
AMF1	Phenanthrene+ pyrene	1.29 (0.08)a	3.05 (0.16)a	0.47 (0.12)a	16.26 (7.30)a
AMF2	Phenanthrene+ pyrene	1.21 (0.23)a	3.86 (1.36)ab	0.53 (0.18)a	11.54 (4.73)ab
AMF1+AMF2	Phenanthrene+ pyrene	0.85 (0.09)bc	3.66 (1.05)a	0.58 (0.11)a	8.87 (4.83)b

Note: Data in brackets are standard deviations (SD). AMF means arbuscular mycorrhizal fungi; "—" means not measured. Values in the same columns followed by the same letter are not significantly different ($p<0.05$).

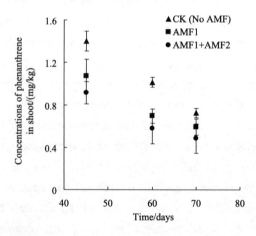

Figure 9-1 Concentrations of phenanthrene in shoots of alfalfa grown in phenanthrene-contaminated soils as a function of uptake time. Error bars are standard deviations

Enhanced PAH uptake by alfalfa roots in the presence of AMF could be attributed to increased PAH partitioning in roots by AMF colonization. Recent studies have shown that hydroponic organic contaminants in soils enter plants primarily via a passive process. This transport process can be treated as a series of contaminant partitions, comprising partitions from soil to soil pore water, from soil water to plant root, and from xylem water to shoot (Li et al., 2005; Gao and Collins, 2009; Gao et al., 2010c). The partitioning of organic chemicals between water and root is the first and predominant step in determining the process of uptake of organic compounds by plants. Extension of mycorrizal hyphae greatly increases the surface area of root-soil contact, and the length of hyphae could be several orders of magnitude greater than that of plant roots (Khan et al., 2000). It might be speculated that mycorrhizal colonization would enhance the partitioning of PAH between root and soil pore water, resulting in increased root concentrations after AMF inoculation. This will be elucidated in "9.2".

The presence of the co-contaminant (pyrene) significantly influenced uptake of phenanthrene by mycorrhizal plants. As shown in Table 9-2, concentrations of phenanthrene in mycorrhizal roots and shoots were clearly much higher in phenanthrene and pyrene-spiked versus single phenanthrene-spiked soils. In our previous work, plant concentrations of PAHs generally increased as PAH concentrations in the soil environment increased (Gao and Ling, 2006). This is also supported by results of other studies (Sung et al., 2001). The observed residual concentrations of phenanthrene in soils after ~70 days were given in Figure 9-2. As seen, the presence of a co-contaminant (pyrene) led to higher residual concentrations of phenanthrene in soils, and consequently resulted in higher concentrations of phenanthrene in the plant body.

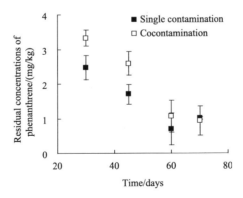

Figure 9-2　Residual concentrations of phenanthrene in soils contaminated by single phenanthrene or phenanthrene with cocontaminant of pyrene. Error bars are standard deviations

The values for SCF and RCF for uptake of phenanthrene by alfalfa with AMF inoculation are provided in Figure 9-3. RCFs for all treatments were 1.48~6.16, which is generally much higher than the corresponding SCFs (0.49~3.11). The presence of a co-contaminant (pyrene) significantly enhanced RCF and SCF values for alfalfa uptake of phenanthrene from spiked soils. For instance, RCFs for phenanthrene-spiked soils with AMF1, AMF2 and both were 1.49, 0.99 and 1.49, respectively, whereas RCFs for phenanthrene and pyrene-spiked soils with these AMF(s) were 6.05, 6.16 and 4.10, respectively. A similar trend was found for SCFs of phenanthrene uptake by alfalfa.

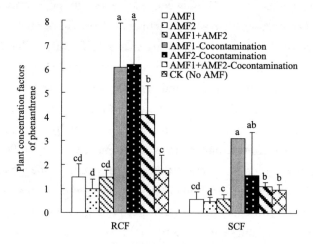

Figure 9-3   Plant concentration factors of phenanthrene for various treatments. RCF and SCF are root concentration factor and shoot concentration factor, respectively. Error bars are standard deviations. Values above columns for RCF or SCF followed by the same letter are not significantly different ($p<0.05$)

In summary, utilizing a greenhouse pot experiment, a novel findling was observed that the colonization of *Glomus mosseae* and *G. etunicatum* caused increased accumulation of phenanthrene and/or pyrene in roots but a decrease in shoots of alfalfa (*Medicago sativa* L.). Results highlight the importance of AMF inoculation on plant uptake of organic contaminants in soil environments.

9.2   Arbuscular mycorrhizal hyphae contribute to PAH uptake by plant

AMF colonization results in the formation of an abundance of extraradical hyphae.

These hyphae can be 5~50 m in length per gram of soil, which is several orders of magnitude longer than plant roots (Khan et al., 2000). With their small diameters (2~15 μm), AMF hyphae have access to sites, such as fine soil pores, that are unavailable to plant roots. AMF hyphae can concentrate and transfer heavy metals from soil to plant bodies (Burkert and Robson, 1994; Chen et al., 2003). However, very little information is available on the uptake of organic contaminants in soils or soil fine pores by AMF hyphae and their translocation to plant roots. The contributions of AMF hyphae to root uptake of organic chemicals-including PAHs-is still under investigation.

Here, we seek to investigate the effects of AMF hyphae on the uptake of fluorene and phenanthrene, as representative PAHs, from soils by roots of ryegrass (*Lolium multiflorum* Lam.) using three-compartment systems (Gao et al., 2010c). Soil and its treatment was the same as that given in "9.1". The final concentrations of fluorene and phenanthrene in the treated soils were 79.52 mg/kg and 72.35 mg/kg (on a dry weight basis), respectively. *Glomus mosseae* (AMF1; BGC GD01A) and *G. etunicatum* (AMF2; BGC HUN02C) were experimented. Results of this work provide insight into the mechanisms of mycorrhizal root uptake of organic contaminants in soils.

9.2.1 Three-compartment systems

Three-compartment systems were used in a greenhouse (Figure 9-4). Each system was divided by metal mesh into three parts (A, B, and C) from top to bottom. Compartments A and B were separated by a 1 mm mesh through which roots could pass. Compartments B and C were separated by a 30 μm mesh through which roots could not pass but AM hyphae could. Compartment C was filled with the spiked soils, whereas compartments A and B were packed with unspiked clean soils. AMFs were inoculated into the compartment B soils; thus, mycelia were born in compartment B, and mycorrhizal hyphae passed through the 30 μm mesh and extended into compartment C. Non-mycorrhizal treatments received the same amount of sterilized inoculum in B compartment. Pregerminated ryegrass seeds were sown in compartment A of each system. The seedlings were thinned 7~10 days after emergence, leaving eight ryegrass plants per pot. The treated pots were located randomly in the greenhouse at 25~30℃ during daytime and at 20~25℃ during night and moved every 4 days. Soils and plants were destructively sampled 30 days, 45 days, 60 days and 70 days after sowing. Plant shoots and roots separated from soils were washed with distilled water, and were prepared for PAH detection.

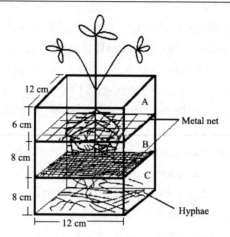

Figure 9-4 Three-compartment cultivation system (Compartments are numbered A, B and C)

9.2.2 Mycorrhizal root colonization and plant biomass

The low AMF colonization rate of the gramineous pasture root was reported in literature. Joner et al. (2001) reported that colonization by *Glomus mosseae* of ryegrass in soils spiked with anthracene, chrysene, and dibenz[a,h]anthracene was 0.5%~5.0%. In this work, mycorrhizal root colonization in the plant growth duration of 30~70 days is shown in Figure 9-5. The colonization rate of the inoculated ryegrass roots generally increased with time, and 17.8% and 16.2% of root colonization were observed for plants inoculated with AMF1 and AMF2 after 70 days, respectively. By contrast, roots of the uninoculated ryegrass remained undetectable.

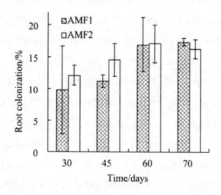

Figure 9-5 Mycorrhizal colonization rates of ryegrass roots after 30~70 days. Error bars are standard deviations

Biomass of the uninoculated, AMF1-inoculated and AMF2-inoculated ryegrass roots were 1.02~3.11 g/plant, 1.32~2.75 g/plant and 1.57~2.47 g/plant, respectively. Similarly, the respective shoot biomass of ryegrass for these treatments was 2.60~7.41 g/plant, 2.38~6.00 g/plant and 2.42~5.66 g/plant in 30~70 days. Further statistical analysis demonstrates that mycorrhizal colonization made no significant difference ($p > 0.05$) to shoot or root yields in the ryegrass growth duration.

9.2.3 Concentrations of PAHs in mycorrhizal roots

The concentrations of fluorene and phenanthrene in the mycorrhizal roots of ryegrass grown in compartments A and B are shown in Table 9-3. After 30~70 days, the concentrations of fluorene in AMF1-colonized roots of compartments A and B were 1.94~62.6 mg/kg and 13.4~44.8 mg/kg, whereas those in AMF2-colonized roots were 0~23.8 mg/kg and 1.28~20.1 mg/kg, respectively. Fluorene concentrations in mycorrhizal roots generally increased initially and then decreased irrespective of compartment. This phenomenon has also been observed in hydroponic experiments. Gao et al. (2006b) investigated the temporal uptake of phenanthrene and pyrene in

Table 9-3 Concentrations (mg/kg, on a dry weight basis) of fluorene and phenanthrene in mycorrhizal root of *Lolium multiflorum* Lam.

PAHs	Arbuscular mycorrhizal fungi	Compartment	Time/day			
			30	45	60	70
Fluorene	AMF1	A	1.94 (0.50)	15.2 (2.10)	62.6 (45.0)	21.8 (10.9)
		B	13.7 (2.06)	13.4 (1.40)	44.8 (9.88)	15.1 (0.13)
	AMF2	A	—	23.8 (2.16)	14.1 (0.74)	—
		B	1.28 (1.10)	13.0 (4.25)	20.1 (3.20)	6.73 (2.67)
	CK	A	—	0.33 (0.10)	0.21 (0.12)	—
		B	—	1.01 (0.24)	1.18 (0.44)	0.23 (0.12)
Phenanthrene	AMF1	A	1.50 (1.35)	0.91 (0.52)	2.86 (1.44)	8.76 (2.63)
		B	0.80 (0.53)	8.87 (1.13)	3.64 (3.40)	0.45 (0.36)
	AMF2	A	1.79 (1.62)	0.93 (0.84)	3.94 (1.15)	11.9 (4.59)
		B	3.86 (1.37)	4.17 (1.25)	4.85 (0.56)	0.27 (0.23)
	CK	A	—	—	0.09 (0.01)	—
		B	—	0.11 (0.01)	0.16 (0.08)	—

Note: Data in bracket are standard deviation (SD); AMF means arbuscular mycorrhizal fungi; CK means nonmycorrhizal control treatments; "—" means undetectable.

solution by ryegrass and red clover. They observed an initial rapid uptake phase followed by a gradual decrease in uptake of tested PAHs based on the plant concentrations and the accumulated amounts of these compounds. Similar results for plant uptake of pesticides have also been reported (Li et al., 2002). Here, the apparent decrease in root concentration of fluorene after the initial increase might be due to dilution by plant growth, metabolism of PAHs in plant tissues, and translocation to other plant parts.

Mycorrhizal root concentrations of phenanthrene were also detected in compartments A and B (Table 9-3). After 30~70 days, concentrations of phenanthrene in roots with AMF1 colonization were 0.91~8.76 mg/kg and 0.45~8.87 mg/kg in compartments A and B, whereas those in roots with AMF2 colonization were 0.93~11.9 mg/kg and 0.27~4.85 mg/kg, respectively. Similar to fluorene, phenanthrene concentrations in mycorrhizal roots increased initially and decreased thereafter in compartment B as a function of uptake time. However, mycorrhizal root concentrations of phenanthrene in compartment A generally increased straightly in 30~70 days. As reported, PAH transportation from root to shoot accompanied with the transpiration stream in plant (Gao and Collins, 2009). Because of phenanthrene's higher molecular weight and lower aqueous solubility, larger volumes of the transpiration stream and longer time are needed for phenanthrene to reach its maximum concentration in ryegrass shoots, resulting in the straight increment of shoot concentrations of phenanthrene in 30~70 days.

The soils in compartments A and B were unspiked and free of PAHs. Roots in these two compartments could not extend to the PAH-spiked compartment because of the 30-μm sieve separating compartments B and C. This was confirmed by the observation that no roots were found in compartment C. However, mycorrhizal hyphae extended through the 30 μm sieve to compartment C. The control treatments without AMF inoculation revealed that concentrations of PAHs in non-mycorrhizal roots of compartments A and B were much lower or undetectable (Table 9-3). These results indicate that the PAHs found in the mycorrhizal roots of ryegrass in compartments A and B were predominantly derived from their uptake by mycorrhizal hyphae in C and their translocation to roots in B and then A.

9.2.4 Partition coefficients of PAHs by arbuscular mycorrhizal hyphae

Studies in recent decades have focused primarily on the uptake and distribution of organic contaminants, including PAHs, in plants (Wang and Jones, 1994; Gao and Zhu,

2004; Wild et al., 2005). However, little information is available on the contributions of mycorrhizal hyphae to the uptake of PAHs by plants. The most novel findings of the present study are the observations of uptake and translocation of PAHs by AMF hyphae using three-compartment systems. Our results proved AMF hyphae could take up fluorene and phenanthrene from the soil environment and transport them to plant roots.

Anthropogenic organic contaminants in soils enter plants primarily via a passive process; this process can be treated as a series of contaminant partitions, including partitions from soil to soil pore water, from soil water to plant root, and from xylem water to shoot (Chiou et al., 2001; Li et al., 2005; Collins et al., 2006). The partitioning of organic chemicals between water and root is the first step in determining the process of organic compound uptake by plants (Li et al., 2005; Gao et al., 2008). Similarly, the uptake of PAHs by AMF hyphae in this investigation could also be considered as having occurred via a series of partitions, and the partitioning of PAHs between the soil solution and the hyphae was the key step dominating the process of hyphal uptake of PAHs from the soil environment.

As such, the partitioning of fluorene and phenanthrene between water solution and AMF hyphae was examined experimentally using a batch equilibration technique (Gao et al., 2010c). Regression was performed on the amounts of fluorene and phenanthrene sorbed by hyphae and their equilibrium concentrations in solution. The sorption isotherms of test PAHs by roots and hyphae could be well described using a linear distribution model (Figure 9-6), indicating a partition-dominated process (Gao et al., 2008; Li et al., 2005).

Partition coefficients (K_d, L/kg) of fluorene and phenanthrene between roots or AMF hyphae and water were obtained through linear regression of sorption data and are listed in Table 9-4. K_d is expressed as

$$K_d = Q_{eq} / C_w$$

where Q_{eq} denotes the amount of fluorene or phenanthrene sorbed by roots or AMF hyphae (mg/kg) and C_w is the equilibrium concentration of fluorene or phenanthrene in the aqueous phase (mg/L). Larger K_d values indicate a greater PAH accumulation ability of the sorbent (roots or hyphae; Chiou et al., 2001; Li et al., 2005). As shown in Table 9-4, K_d values for fluorene and phenanthrene in AMF hyphae were very high (8576~10564 L/kg) and were 270%~356% greater than those for plant roots. This indicates that AMF hyphae have higher partitioning abilities for lipophilic organic

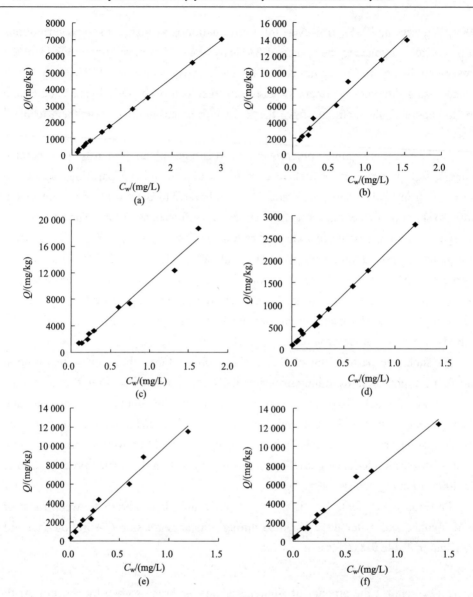

Figure 9-6 Sorption of fluorene and phenanthrene by root and arbuscular mycorrhizal hyphae. C_w-equilibrium concentration of PAH in solution; Q-sorption amount of PAH. (a) Sorption of fluorene by root; (b) Sorption of fluorene by AMF1 hyphae; (c) Sorption of fluorene by AMF2 hyphae; (d) Sorption of phenanthrene by root; (e) Sorption of phenanthrene by AMF1 hyphae; (f) Sorption of phenanthrene by AMF2 hyphae

chemicals such as PAHs and can readily concentrate the tested PAHs from the soil solution in compartment C, resulting in the greater uptake ability of AMF hyphae for fluorene and phenanthrene. This is consistent with and further confirmed the mycorrhizal root uptake observations.

Table 9-4 Some parameters for PAH sorption by tested root and arbuscular mycorrhizal hyphae

Sorbents	phenanthrene		fluorene	
	R^2	K_d/(L/kg)	R^2	K_d/(L/kg)
root	0.9958	2275	0.9988	2316
Hyphae of AMF1	0.9736	9499	0.9806	8576
Hyphae of AMF2	0.9858	8862	0.9567	10564

Note: R^2 is correlation coefficient; K_d is the partition coefficient.

9.2.5 Translocation of PAHs by arbuscular mycorrhizal hyphae

This study demonstrated the uptake of PAHs by arbuscular mycorrhizal hyphae and their translocation from hyphae to host plant roots. To evaluate the translocation ability of different PAHs by fungal hyphae, we introduced a parameter, TF (translocation factor). The value of TF was determined as follows

$$TF = C_{Root} / C_{Soil}$$

where C_{Root} denotes the PAH concentration in mycorrhizal roots in compartment B (mg/kg) and C_{Soil} is the soil concentration of PAHs in compartment C (mg/kg). A larger TF value indicates that more PAHs are translocated by hyphae from soil to root.

The TF values of fluorene and phenanthrene for AMF hyphae are shown in Figure 9-7. The TF of fluorene for AMF1 and AMF2 hyphae from day 30 to day 70 ranged from 0.95 to 9.95 and from 1.33 to 5.85, whereas that of phenanthrene was from 0.55 to 3.07 and from 0.53 to 3.60, respectively. Generally, TF values first increased and then decreased from day 30 to day 70. The TF values of fluorene were 62.2%~1020% greater than those of phenanthrene.

AMF colonization results in the formation of an abundance of extraradical hyphae (5~50 m hypha per gram of soil). Because of their small diameter (2~15 μm), the fine soil pores interwoven by AMF hyphae have access to sites unavailable to roots (Khan et al., 2000; Hart and Trevors, 2005). In theory, this extends the contact area of AM

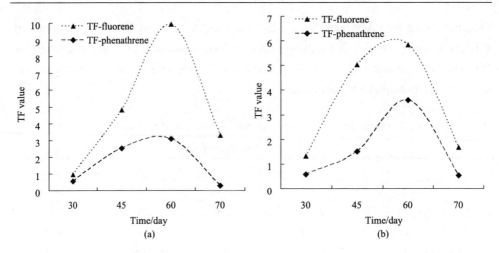

Figure 9-7 Translocation factors (TFs) of fluorene and phenanthrene by arbuscular mycorrhizal hyphae.

(a) AMF1, (b) AMF2

roots with soils. AMF hyphae reportedly can transport water and small molecules (Trotta et al., 2006). Chen et al. (2003) reported that AMF hyphae can absorb zinc directly from the soil and transfer it to plant roots. Using culture systems that separated extraradical hyphae from roots, Burkert and Robson (1994) observed that AMF hyphae can accumulate and translocate ^{65}Zn to a degree that may differ between species. Transportation of unnecessary elements by hyphae is possibly a selective mechanism. Direct evidence of the translocation of hydrophobic organic chemicals by mycorrhizal hyphae is scarce. However, the present study indirectly demonstrates the transportation of PAHs by AMF hyphae from soil to plant roots.

PAHs can be transferred from roots to shoots along the transpiration stream (Gao and Zhu, 2004; Gao and Collins, 2009). In arbuscular mycorrhizal systems, the transpiration stream flows additionally from AMF hyphae to roots and then to shoots. PAHs dissolved in soil pore water are absorbed by the hyphae and are then partitioned to hyphae and plants, resulting in the translocation of these chemicals by mycorrhizal hyphae. In theory, organic chemicals with low molecular weights and higher water solubility are more readily dissolved and therefore accompany the transpiration stream. This was supported here by the higher TF values of fluorene versus phenanthrene for AMF hyphae. Fluorene and phenanthrene both contain three fused benzene rings, but phenanthrene is larger and more hydrophobic, with a larger octanol-water partition coefficient (K_{ow}) than fluorene. By contrast, the water solubility (S_w) of fluorene is higher than that of phenanthrene. These characteristics of fluorene provide for more opportunities for translocation from soil to root by mycorrhizal hyphae accompanied by water. These results suggest that smaller PAHs with higher S_w were more readily translocated by AMF hyphae.

In summary, the present study firstly demonstrates the uptake and translocation of PAHs in soils by AMF hyphae using three-compartment systems. Although ryegrass roots were grown in unspiked clean soil, abundant mycorrhizal hyphae extended into PAH-spiked soil and took in and transported PAHs to plants, resulting in high concentrations of PAHs in roots. This study provides new views of the AMF hyphae-mediated uptake of organic soil contaminants by plants.

Chapter 10 Utilizing PAH-degrading endophytic bacteria to reduce the plant PAH contamination

Plant-associated microorganisms play key roles in PAH uptake by plants (Viñas et al., 2005; Sheng et al., 2008). These microbes can participate in PAH-degrading processes (Yuan et al., 2011). The inoculation of arbuscular mycorrhizal fungi (AMF) has proven to be effective in reducing the PAH concentrations in soils (Gao et al., 2011c). PAH-degrading bacteria have also been isolated and utilized in the degradation of PAHs. However, the reported PAH-degrading bacteria have been overwhelmingly isolated from soil, sediment, and sludge, and little information is available on the enormous potential of endophytes in mediating the uptake of PAHs by plants (Dai et al., 2010).

Endophytic bacteria, defined as a class of microbes that reside within the interior tissues of plants without causing harm to host plants or environments (Harish et al., 2008; Liu et al., 2014), form one of the microbial communities most closely associated with plants. They have established harmonious associations with host plants during including symbiotic, mutualistic, commensalistic, and trophobiotic relationships over a long evolutionary process (Lodewyckx et al., 2002). This assortment of bacteria have a wide variety of functions including the stimulation of plant growth (Shin et al., 2007), the promotion of biological nitrogen fixation (Davison, 1988), the protection of plants from harsh external environments, and the control of pathogen activities (Brooksd et al., 1994). Previous studies have shown that some endophytic bacterial strains have the ability to degrade organic pollutants in plants and soils (Rajkumar et al., 2009; Khan and Doty, 2011; Zhu et al., 2014). Germaine et al. (2006) inoculated pea plants with a 2,4-D-degrading endophytic bacterium (*Pseudomonas putida* VM1450) and found that this strain can internally colonize plants, maintain their growth, and cause a 24%~35% increase in contaminant removal from the plants. Therefore, the use of endophytic bacteria to regulate the metabolism of organic pollutants and to reduce contamination risks in plants would be significantly advantageous (Sheng et al., 2008). However, little literature is available on the utility of PAH-degrading endophytes in reducing PAH contamination in plants.

In this chapter, the distributions of endophytic bacteria in plants grown in soils contaminated with PAHs were investigated with polymerase chain reaction followed by denaturing gradient gel electrophoresis technology (PCR-DGGE) and cultivation methods. PAH-degrading endophytic bacteria were isolated from clover plants grown in PAH-contaminated sites. After being marked with the GFP gene, the colonization and distribution of isolated strains were visualized in plants. The PAH-degradation performance of the endophytic bacteria in plants was evaluated. Investigations provide a novel method of endophyte-aided removal of PAHs from plant bodies.

10.1 Distribution of endophytic bacteria in plants from PAH-contaminated soils

Assessing the diversity, distribution, physiology, and ecology of endophytic bacteria in plants is prerequisite for isolating organic contaminant-degrading endophytic bacteria and using them to eliminate organic pollution in plants (Moore et al., 2006; Tang et al., 2010). There have been some reports regarding endophytic bacterial populations in plants grown in soils polluted with different contaminants (Nogales et al., 2001; Kaplan and Kitts, 2004). Moore et al. (2006) investigated endophytic bacterial populations in poplar trees and found that a number of isolates had the ability to degrade BTEX compounds or to grow in the presence of TCE. Ho et al. (2009) isolated 188 endophytic strains from three plants and found that among these strains, 29 not only grew well in the presence of naphthalene, catechol, and phenol but were also able to utilize the pollutant as a sole carbon source for growth. However, to our knowledge, little information is available about endophytic bacterial populations in plants from PAH-contaminated sites.

Here, utilizing PCR-DGGE technology combined with culture-dependent methods, we seek to investigate the distribution and diversity of endophytic bacteria in two plants (*Alopecurus aequalis* Sobol and *Oxalis corniculata* L.) that are common in China and also the dominant plants in a PAH-contaminated field (Peng et al., 2013, 2015). Amur foxtail (*Alopecurus aequalis* Sobol), creeping oxalis (*Oxalis corniculata* L.), and soil samples were obtained from an aromatics factory in Nanjing. Samples were collected from three stations (A, Z, and Q) at different distances from the aromatics factory. The physicochemical characteristics of the sampled soils were as follows: pH 5.87, 13.0% sand, 60.7% silt, 26.3% clay and 1.36% organic matter. The total PAH contents of soils from stations A, Z, and Q were 178 mg/kg, 139 mg/kg and 89.5 mg/kg, respectively. Table 1 shows the concentrations of each PAH in soils. The plant samples were removed from the soil. Some of the soil and plant samples were freeze dried immediately for the determination of PAH contents. The plant samples were rinsed with deionized water and subsequently surface sterilized. Total DNA extraction, DGGE analysis, DNA sequence analysis, and isolation of cultivable endophytic bacteria in plants were performed in details (Peng et al., 2013). This is an indispensable precondition for the future isolation of functional endophytes to aid in the reduction of plant PAH pollution risk.

Table 10-1 Concentrations of PAHs in soils

PAHs	Abbreviation	The concentrations of PAHs in soil/(mg/kg dry weight)		
		A	Z	Q
Naphthalene	NAP	34.3±3.80a	25.5±7.30a	25.7±4.19a
Acenaphthylene	ANE	42.8±1.77a	39.9±0.48ab	36.7±0.64b
Phenanthrene	PHE	8.62±0.67a	6.45±0.58a	3.36±0.78b
Anthracene	ANT	72.1±4.71a	56.6±4.34a	16.1±7.92b
fluorine	FLU	0.89±0.02a	0.72±0.06b	0.73±1.44b
acenaphthylene	ANY	2.09±0.05a	1.01±0.04b	1.04±0.06b
Pyrene	PYR	1.68±0.02a	0.94±0.01b	0.91±0.03b
fluoranthene	FLA	3.19±0.18a	2.14±0.09ab	0.82±0.34b
chrysene	CHR	6.62±0.41a	3.20±0.04b	1.22±0.49c
benz[a] anthracene	BaA	0.94±0.01a	0.51±0.17b	0.52±0.01b
benzo[k]fluoranthene	BbF	1.77±2.87a	0.86±4.41b	0.69±1.97b
benzo[ghi]perylene	BghiP	3.50±0.03a	1.64±0.04b	1.66±0.12b
∑PAHs		178±4.87a	139±4.20b	89.5±5.79c

Note: different letters in the same row indicate significant differences ($p < 0.05$).

10.1.1　PAH concentrations in plants from PAH-contaminated soils

In this study, 12 types of PAHs that have been designated by the US Environmental Protection Agency as priority pollutants were detected in soil and plant samples, including NAP, ANE, PHE, ANT, FLU, ANY, PYR, FLA, CHR, BaA, BbF and BghiP. As shown in Table 10-1, soil samples obtained from each sampling station contained different concentrations of PAHs, and the total PAH contents of soils from A, Z, and Q stations were 178 mg/kg, 139 mg/kg and 89.5 mg/kg, respectively. ANOVA revealed significant differences among the total PAH concentrations at the three sampling stations. ANE and ANT were the main components in all three soil samples, accounting for 23.9%~40.9% and 18.0%~40.6% of the total concentrations, respectively.

The 12 types of PAHs detected in rhizosphere soils were also determined in the roots and shoots of the two plants (Tables 10-2 and 10-3), and the overwhelming majority of PAHs accumulated in plant roots. For example, at position A, PAHs were enriched to a concentration of 217 mg/kg in the roots of *A. aequalis*; however, in the shoots, the total PAH concentration was only 88.0 mg/kg. Furthermore, for both plants, as soil PAH concentrations increased, the PAH concentrations in the plants also increased. As shown in Tables 10-2 and 10-3, the PAH concentrations in the roots of plants from station A were significantly higher than those in plants from the other two stations. Two-ring to three-ring PAHs were the main detectable pollutants in plant samples, with proportions of 94.9%~96.2% in *A. aequalis* and 87.8%~94.2% in *O. corniculata* grown in differentially contaminated soils. Conversely, four-ring to six-ring PAHs accounted for only minor proportions of the total PAHs in the two plants (approximately 3.8%~5.1% and 5.8%~12.2%, respectively).

Table 10-2 Concentrations of PAHs in *Alopecurus aequalis*

PAHs	Root/(mg/kg dry weight of plant)			Shoot/(mg/kg dry weight of plant)		
	A	Z	Q	A	Z	Q
NAP	120±5.43a	45.5±8.11b	59.9±21.1ab	42.8±5.30a	26.9±5.34a	20.2±7.82a
ANE	33.9±7.81a	28.8±6.26a	24.9±7.80a	15.9±3.36a	12.5±0.83a	9.32±1.85a
PHE	4.85±0.21a	4.12±0.01a	3.7±1.31a	2.31±0.28a	2.25±0.03a	1.98±0.71a
ANT	49.0±2.73a	44.3±6.64a	30.5±9.02a	22.6±3.80a	20.7±1.39a	23.2±3.93a
FLU	0.44±0.01a	0.39±0.01a	0.36±0.13a	0.21±0.12a	0.22±0.01a	0.19±0.06a
ANY	0.73±0.04a	0.63±0.05a	0.56±0.22a	0.39±0.10a	0.45±0.10a	0.34±0.07a
PYR	0.62±0.04a	0.59±0.07a	0.48±0.06a	0.30±0.02a	0.27±0.01a	0.30±0.01a
FLA	1.72±0.04a	1.54±0.22a	1.28±0.24a	0.82±0.12a	0.75±0.07a	0.81±0.04a
CHR	3.78±0.37a	2.90±0.15ab	1.81±0.57b	1.77±0.05a	1.66±0.34a	1.61±0.08a
BaA	0.32±0.01a	0.29±0.03a	0.25±0.02a	0.16±0.01a	0.17±0.00a	0.16±0.00a
BbF	0.54±0.10a	0.51±0.04a	0.43±0.13a	0.31±0.03a	0.34±0.02a	0.31±0.04a
BghiP	0.77±0.01a	0.71±0.04a	0.60±0.10a	0.38±0.01a	0.36±0.00a	0.38±0.01a
$\sum$PAHs	217±11.7a	130±12.0b	124±2.30b	88.0±3.97a	66.5±2.41b	58.8±4.08c
CF_{PAHs}	1.21	0.93	1.39	0.49	0.48	0.66

Note: different letters in the same row indicate significant differences ($p<0.05$); CF_{PAHs} means the plant concentration factors of total PAHs.

Table 10-3　Concentrations of PAHs in *Oxalis corniculata*

PAHs	Root/(mg/kg dry weight of plant)			Shoot/(mg/kg dry weight of plant)		
	A	Z	Q	A	Z	Q
NAP	109±6.53a	64.1± 12.0b	63.3±19.6b	49.2± 2.78a	45.8±15.76a	19.7±10.9b
ANE	83.2±3.09a	40.2±2.14b	38.5±3.56b	30.2±3.71a	32.3±10.61a	12.9±3.71b
PHE	18.2±2.83a	9.28±0.34b	8.4±2.47b	5.25± 0.32a	4.53±1.53ab	2.26±0.54b
ANT	180.±10.12a	85.8±0.42b	68.5±9.88c	47.2±1.41a	41.3±3.13a	19.6 ±4.13b
FLU	1.69±0.06a	0.78±0.02b	0.81±0.21b	0.48±0.04a	0.48±0.19a	0.29±0.02a
ANY	2.22±0.40a	1.16±0.09b	1.07± 0.30b	0.71±0.08a	0.75±0.22a	0.40±0.01a
PYR	2.32±0.88a	1.12±0.07b	1.14±0.27b	0.68±0.02a	0.61± 0.20a	0.36±0.02b
FLA	6.01±1.12 a	2.97±0.28b	1.67±1.26c	1.75±0.08a	1.44±0.63ab	0.64±0.27b
CHR	50.0±6.58a	6.18±3.83b	5.83±1.22b	3.58±0.30a	2.78±1.36a	1.45±0.96a
BaA	1.22±0.24a	0.60±0.09b	0.60±0.17b	0.37±0.02a	0.31±0.09a	0.21±0.01a
BbF	2.28±0.49a	1.13±0.15b	0.60±0.25c	0.69±0.06a	0.50±0.04a	0.44±0.06a
BghiP	3.21±0.67a	1.58±0.08b	1.48±0.35b	0.89±0.03a	0.77±0.09ab	0.45±0.03b
∑PAHs	459±10.6a	215±14.2b	192±23.0b	141±5.74a	132±12.4a	58.7±9.34b
CF_{PAHs}	2.57	1.54	2.15	0.79	0.94	0.66

Note: different letters in the same row indicate significant differences (p <0.05); CF_{PAHs} means the plant concentration factors of total PAHs.

However, there were obvious differences in the absorption and accumulation of PAHs between the two plants. *O. corniculata* was better able to accumulate PAHs compared with *A. aequalis*. For example, the PAH concentrations in *O. corniculata* grown at positions A, Z and Q were 251 mg/kg, 346 mg/kg and 600 mg/kg, respectively, which were significantly higher than those in *A. aequalis* (184 mg/kg, 197 mg/kg, and 304 mg/kg, respectively at positions A, Z, and Q). Additionally, the concentration of NAP in *A. aequalis* was the highest among all detected PAHs, accounting for 36.8%~53.4% of the total PAH content of the plants from the different stations. In contrast, the concentration of ANT was highest in *O. corniculata*, accounting for 60.9%~62.5% of the total PAH content.

The total PAH concentration in *O. corniculata* was significantly higher than that in *A. aequalis*. This may be attributable to the different growing seasons (*O. corniculata* is a perennial herb whereas *A. aequalis* is an annual herb) or the different lipid contents of the two plants. Chiou et al. (2001) built a partition-limited model indicating that water-insoluble contaminants, even in small amounts, were the major compounds present in the plant-lipid phase. Zhu and Gao (2004) also confirmed a significant positive correlation between the root concentration of phenanthrene and root lipid content. Interestingly, both *A. aequalis* and *O. corniculata* absorbed few high molecular weight PAHs with larger numbers of benzene rings (pentacyclic and hexacyclic), which could be due to the poor bioavailability of these types of PAHs (Juhasz and Naidu, 2000).

10.1.2 Endophytic bacterial community in PAH-contaminated plants

The PCR-DGGE profiles are shown in Figure 10-1. All plant samples from the three sites contained band 1 (*Pseudomonas* sp.), indicating that one of the dominant bacteria was the same in both plants at different levels of PAH contamination. However, there were also some discrepancies between different plant tissues and different places. For example, band 2 (uncultured bacterium clone) was only existed at site Q, and band 3 (*Nesterenkonia* sp.) was only existed at sites Q and Z. Pollution intensity could significantly influence the diversity of the endophytic bacterial community. As shown in Figure 10-2, the plants that grew in lightly polluted soils consistently showed the highest diversity.

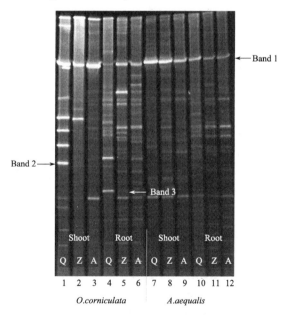

Figure 10-1 Representative DGGE for PCR-amplified 16S rDNA V3 fragments from endophytic bacteria in *Alopecurus aequalis* and *Oxalis corniculata*. The genus of the band marked in the figure: Band 1-*Pseudomonas* sp.; Band 2 - uncultured bacterium clone; Band 3- *Nesterenkonia* sp.

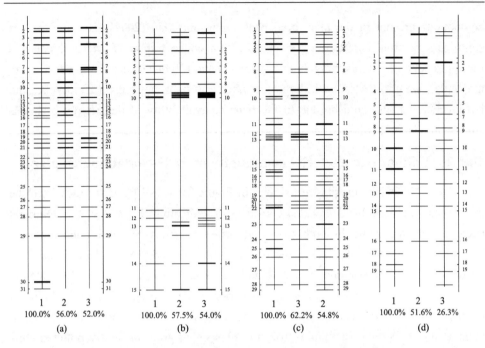

Figure 10-2 The similarity of endophytic bacterial community in roots of *Alopecurus aequalis* (a), shoots of *Alopecurus aequalis* (b), roots of *Oxalis corniculata*(c) and shoots of *Oxalis corniculata* (d).1-Q pollution level (defined as 100%), 2-Z pollution level, 3-A pollution level

Nearly all the bands found in the DGGE gel were sequenced, and after removing the 16S rDNA and 18S rDNA of mitochondria and chloroplasts, we managed to obtain exactly 77 different bacterial sequences, among which 56 were isolated from *A. aequalis* (root, 37; shoot, 22; both, 3) and 32 were obtained from *O. corniculata* (root, 27; shoot, 12; both, 4). Overall, 44.1% of the sequences isolated from *A. aequalis* were derived from uncultured bacteria, 8.47% were derived from *Pseudomonas* sp. and 6.78% were derived from *Halomonas* sp. The remaining 40.7% were derived from approximately 20 different genera of bacteria. In *O. corniculata,* the highest percentage of sequences (30.6%) was also derived from uncultured bacteria, followed by *Pseudomonas* sp. (19.4%) and *Enterobacter* sp. (8.33%). The remaining 41.7% of sequences were derived from 15 different genera of bacteria.

Phylogenetic analysis (Figure 10-3) of the endophytic bacterial community performed using the results of the DGGE gel indicated that these bacteria belong to four phyla (Firmicutes, Proteobacteria, Actinobacteria and Bacteroidetes), seven classes(*Bacilli, Clostridia,α-proteobacteria, β-proteobacteria, γ-proteobacteria, Acti-*

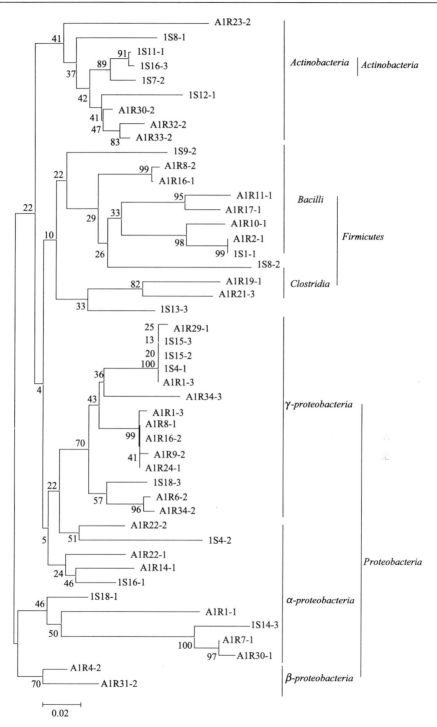

(a)

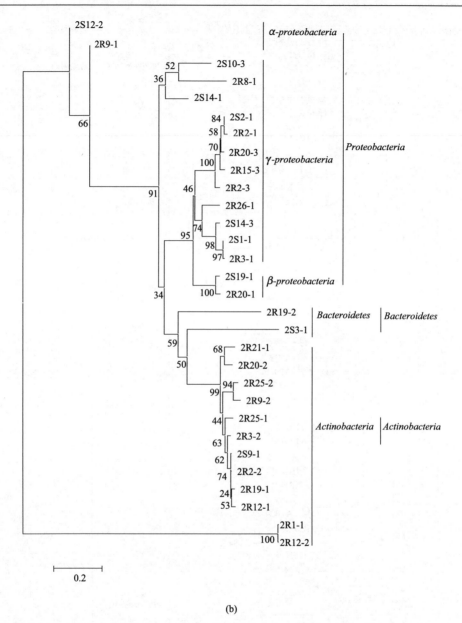

(b)

Figure 10-3　Phylogenetic trees by 16S rDNA V3 sequence analysis of endophytic bacterial community of *Alopecurus aequalis* (a) and *Oxalis corniculata* (b). The codes with characters and numbers indicate the isolated DGGE bands. Bootstrap values are shown for each node in a bootstrap analysis of 1000 replicates *nobacteria, Bacteroidetes*), and over 30 families. Most of the isolates from *A. aequalis* belonged to Proteobacteria and Firmicutes, and the remaining belonged to Actinobacteria. Proteobacteria also composed a large percentage of the bacteria isolated from *O. corniculata*, and the remaining bacteria from *O. corniculata* belonged to Bacteroidetes and Actinobacteria.

Compared with culture-dependent method, using PCR-DGGE technology we found that bacterial community profiles showed extensive variability depending on the plant origin, tissue, and PAH levels in soil. To further explore this dissimilarity, the Shannon diversity index was calculated for each sample using Quantity One and Bio-Dap software. The results show that from Q to A, the Shannon diversity indices of *A. Aequalis* were 3.38, 3.2 and 2.93 in roots and 2.55, 3.03 and 2.74 in shoots, respectively. In *O. Corniculata*, the Shannon diversity indices were 3.34, 3.27 and 3.29 in roots and 2.78, 2.66 and 2.73 in shoots. The Shannon index includes two basic components: abundance and evenness of the species present (Kapley et al., 2007), and this result indicated that different plant samples and pollution levels could lead to different bacterial distributions and levels of diversity. Qing et al. (2006) and Robson et al. (2010) also obtained a similar conclusion in their studies. Through phylogenetic analysis (Figure 10-3), we can see that the most prominent groups in the two plants were both related to *Proteobacteria*, which agrees with other studies (Chelius and Triplett, 2001; Sun et al., 2008).

10.1.3 Cultivable endophytic bacterial populations in PAH-contaminated plants

A total of 68 species of endophytic bacteria were isolated and identified, among which 39 were isolated from *A. aequalis* (root, 27; shoot, 20; both, 7), 28 were isolated from *O. corniculata* (root, 15; shoot, 17; both, 3) and 1 was isolated from both plants. The isolates were identified using 16S rDNA analysis, and these 16S rDNA sequences shared high identities with their most closely related species in the database ($\geqslant 97\%$), with most having identities of 99%~100% with known bacterial species.

The cultivable endophytic bacterial populations of *A. aequalis* and *O. corniculata* grown in PAH-contaminated soils showed low diversity. The 68 endophytic bacterial species were classified into 14 genera belonging to five classes, including *Bacilli*, *α-proteobacteria*, *β-proteobacteria*, *γ-proteobacteria* and *Flavobacteriaceae*. Although more endophytic bacterial species were isolated from *A. aequalis* than from *O. corniculata*, the endophytic bacterial population in *O. corniculata* included more genera than that in *A. aequalis*. In *A. aequalis*, *Bacillus* and *Pseudomonas* spp. constituted a large proportion, accounting for 65.5% and 24.1%, respectively, of the total bacterial population. Meanwhile, members of *Bacillus* spp. accounted for the highest proportion of strains in *O. corniculata*, constituting 43.9% of the total bacterial population, followed by *Pseudomonas* spp. and *Rahnella* spp., each constituting 21.1% of the total population.

Studying endophytic bacterial populations under different pollution levels is a popular means of examining contaminant-degrading bacterial flora (Moore et al., 2006; Khan and Doty, 2011). Siciliano et al. (2001) found that the amount of contaminant-degrading bacterial cells increased with exposure to a contaminated environment. Phillips et al. (2008) analyzed the relationship between the endophytic bacterial community and the capacity for organic pollutant degradation in alfalfa. These authors found that when the dominant population consisted of *Pseudomonas* spp., the ability of plants to metabolize alkane pollutants improved, whereas when the dominant bacterial populations shifted to *Brevundimonas* spp. and *Pseudomonas rhodesiae*, the capacity of the plant to metabolize aromatic hydrocarbons increased. This suggests that the dominant endophytic bacterial taxa isolated from contaminated plants may have the potential to degrade PAHs in plants (Moore et al., 2006). Thus, further studies of the tolerance of isolated endophytic bacterial strains to various PAHs were performed.

10.1.4 Amounts of cultivable endophytic bacteria in PAH-contaminated plants

As shown in Figure 10-4, the cultivable endophytic bacteria detected in the two studied plants were on the order of 10^4 to 10^7 CFU per gram of fresh plant tissue in most cases, and the number of endophytic bacteria in roots was higher than that in shoots. As the PAH concentration increased, the total number of cultivable endophytic bacterial strains was reduced. For example, the number of endophytic bacterial strains in *A. aequalis* from station Q was 251 times and 27 times larger than those from stations A and Z, respectively, but the concentration of PAHs was only 0.5 times and 0.8 times of those from stations A and Z, respectively. Furthermore, as PAH concentrations increased, the total number of endophytic bacterial cells in plant tissues decreased by different degrees, and the range varied more in plant roots compared with shoots. For example, the number of endophytic bacteria in *A. aequalis* roots from station Q was 351 times that of roots from station A, while in shoots, the difference was only 3.23 times greater at station Q compared with station A.

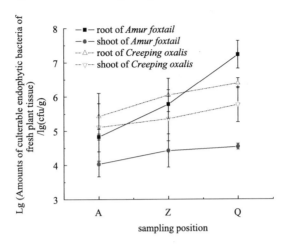

Figure 10-4　Amounts of cultivable endophytic bacterial strains in *Alopecurus aequalis* and *Oxalis corniculata* under different PAH pollution levels

PAH pollution levels also had impacts on the dominant populations of cultivable endophytic bacteria in the two plants (Figure 10-5). As PAH concentrations increased, the proportion of *Bacillus* spp. and *Pseudomonas* spp. showed opposite tendencies in the roots and shoots of *A. aequalis*. For example, under higher PAH concentrations (station A), *Bacillus* spp. accounted for 82.4% of the total endophytic bacteria in roots, and at the other two stations, *Bacillus* spp. accounted for 78.5% (Z) and 51.4% (Q). However, *Bacillus* spp. in the shoots of *A. aequalis* showed the opposite trend, with proportions of 15.4%, 70.8% and 89.4% in samples from stations A, Z, and Q, respectively. Meanwhile, with increasing pollution levels, the proportion of *Pseudomonas* spp. decreased in plant roots and increased in shoots. The dominant populations in *O. corniculata* from different stations were quite different. The dominant genus in roots was *Rahnella* spp. when the pollution level was relatively high (A, Z); however, under a lower level of pollution, the dominant genus became *Pseudomonas* spp..

Although endophytic bacteria have been previously reported in many plants including sugarcane (Mendes et al., 2007), ginseng (Vendan et al., 2010), and aquatic plants (Chen et al., 2012), among others, little information is available regarding endophytic bacteria in plants grown in PAH-contaminated soils. In this study, a total of 68 endophytic bacterial strains were isolated in *A. aequalis* and *O. corniculata* growing

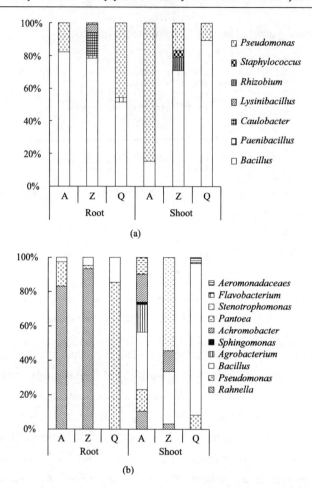

Figure 10-5 Amounts of cultivable endophytic bacterial strains belonging to each genus in *Alopecurus aequalis* (a) and *Oxalis corniculata* (b) under different PAH pollution levels

in PAH-contaminated stations. As known to us, Bacilli are a group of the commonest bacteria associated with plants and take a certain proportion of the cultivable endophytic bacteria in lots of plants (West et al., 2010; Chen et al., 2012). In our study, members of *Bacillus* spp. account for the highest proportion of cultivable endophytic bacterial strains in both *Alopecurus aequalis* Sobol and *Oxalis corniculata* L., followed by *Pseudomonas* spp. Many bacterial strains belonging to these two genera are known to be capable of degrading various types of organic pollutants, which accounts for the frequent occurrence of these strains in organically contaminated soils in relatively high amounts (Moore et al., 2006). In addition, the numbers of endophytic bacteria in the two plants were quite high: $7.84 \times 10^4 \sim 1.70 \times 10^7$ cfu/g and $4.40 \times 10^5 \sim 3.06 \times 10^6$ cfu/g in *A. aequalis* and *O. corniculata*, respectively. In contrast, Garbeva et al. (2001) and Araújo et al. (2002) obtained only $1.0 \times 10^3 \sim 1.0 \times 10^5$ cfu/g and $10^3 \sim 10^4$ cfu/g of endophytic bacterial cells from the potato and citrus plants, respectively. Some endophytic bacterial strains, such as strains AF14, AF19, and CO4, occurred in both the roots and shoots of a given plant, indicating that endophytic bacterial strains can be transported and spread throughout plant systems (McCully, 2001; Compant et al., 2010).

Plant-associated habitats are dynamic environments in which many factors affect the species compositions of microbial communities (Zak et al., 2003). The host plant species is one of the major influencing factors (Sessitsch et al., 2002). For instance, Chen et al. (2012) studied endophytic bacterial species of four aquatic plants, and their results showed that the dominant endophytic bacterial taxa in the four plants were quite different; *Pseudomonas* spp. and *Staphylococcus* spp. were the dominant taxa in *Phragmites communis* and *Potamogeton crispus*, whereas in *Nymphaea tetragona* and *Najas marina*, the dominant taxa were *Aeromonas* spp. and *Bacillus* spp. In this study, at the same pollution level, the dominant endophytic bacteria in *A. aequalis* and *O. corniculata* were also different. Pollution stress within a plant growing area is another major environmental factor influencing endophytic bacterial communities (Phillips et al., 2008). Sobral et al. (2005) found that many endophytic bacteria could be cultivated from soybeans grown in soil to which glyphosate had been previously applied (pre-planting); however, when the glyphosate was enriched, the taxa of the cultivable endophytic bacteria changed. Similar results were observed in this study: The proportion of each endophytic bacterial genus in a given plant was found to vary at different pollution levels. Even among different tissues of one plant, when the total PAH concentration varied, the endophytic bacterial population changed correspondingly (Figure 10-5).

10.2 Inoculating plants with the endophytic bacterium *Pseudomonas* sp. Ph6-*gfp* to reduce phenanthrene contamination

The association of endophytic bacteria with plants may be used to obtain more efficient degradation of organic pollutants and to reduce organic contaminants in plants (Ryan et al., 2008; Germaine et al., 2009). A 2,4-D-degrading endophytic bacterium, *Pseudomonas putida* VM1450, colonized pea plants, maintained plant growth, and increased contaminant removal from the plants by 24%~35% compared with the control (Germaine et al., 2006). While some endophytic bacteria degrade organic pollutants *in vitro*, other bacteria might act only in plants. Hence, the utilization of endophytic bacteria to reduce the risk of organic contamination in plants would have immense advantages (Sheng et al., 2008). However, the interactions between an endophytic bacterium and its plant host that are required for metabolizing PAH are poorly understood.

Here, the phenanthrene-degrading endophytic bacterium *Pseudomonas* sp. Ph6 was isolated from clover (*Trifolium pratense* L.) grown in PAH-contaminated soil (Sun et al., 2014a). After being *gfp*-tagged, Ph6-*gfp* colonization and its performance on PAH uptake by ryegrass (*Lolium multiflorum* Lam.) were systematically investigated (Sun et al., 2015).

10.2.1 Isolation, identification, and *gfp*-labeling of *Pseudomonas* sp. Ph6

Clover (*Trifolium pratense* L.) grown in PAH-contaminated sites around a petrochemical plant in Nanjing, China, was collected for the isolation of phenanthrene-degrading endophytic bacteria. The isolation and identification of endophytic bacterium *Pseudomonas* sp. Ph6 were referred to Sun et al. (2014a). To construct the *gfp*-tagged *Pseudomonas* sp. Ph6 (strain Ph6-*gfp*), plasmid pBBRGFP-45 was transformed into *Pseudomonas* sp. Ph6 by triparental conjugation (Table 10-4). The positive transconjugants were purified, and the presence of *gfp* gene was confirmed by PCR. The expression level of *gfp* was examined using fluorescence microscopy under UV light (Sun et al., 2014a).

Table 10-4 Strains and plasmids for triparental conjugation. Bacteria containing the plasmid pBBRGFP-45 (Km^r, 50 mg/L), bacteria containing the plasmid pRK2013 (Km^r, 50 mg/L), and strain Ph6 (Cm^r, 50 mg/L)

Strain or plasmid	Genotype or phenotype	Source or reference
pBBRGFP-45	pBBR-MCS2 vector with 1.4 kb foreign fragment (*gfp*)	Yu(2007)
pRK2013	mob$^+$, tra$^+$, Km^r, helper plasmid	This laboratory
Pseudomonas sp. Ph6	Phenanthrene degradation	This laboratory

The phenanthrene-degrading endophytic bacterium *Pseudomonas* sp. Ph6 was isolated from collected plants. This strain is a rod-shaped, aerobic, gram-negative bacterium that is motile, with polar flagella. Ph6 was positive for nitrate reductase, negative for indole and gelatin liquefaction, and positive for malate, citric acid, and phenylacetic acid. The glucose test was positive, while mannose and maltose were negative. The 16S rRNA gene sequence of strain Ph6 showed 99% identity to *Pseudomonas* sp.. The GenBank accession number of the 16S rRNA gene sequence of strain Ph6 is KF741207. On the basis of its morphology, physiology, and 16S rRNA gene sequence analysis, Ph6 was identified as a *Pseudomonas* sp. bacterium. The cellular morphology of strain Ph6 is shown in Figure 10-6.

The plasmid pBBRGFP-45 was successfully transformed into endophytic bacterium *Pseudomonas* sp. Ph6, and strain Ph6 tagged with GFP gene (Ph6-*gfp*) showed stable *gfp* activity. The presence of the plasmid pBBRGFP-45 in the transconjugants was confirmed by PCR (data not shown). The cells carrying the plasmid were detected by fluorescence microscopy (Figure 10-7). The morphologically tagged strain *Pseudomonas* sp. Ph6-*gfp* was highly similar to the wild strain *Pseudomonas* sp. Ph6, and the growth rate of Ph6-*gfp* in liquid LB was also similar to the wild type (Figure 10-8). This result indicates that the presence of the plasmid did not interfere with the normal growth and activity of Ph6.

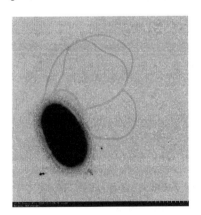 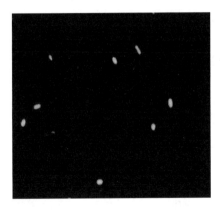

Figure 10-6 Transmission electron micrograph of strain Ph6 isolated from clover (*Trifolium pratense* L.) (×4.0 K Zoom-1 HC-1 80 kV)

Figure 10-7 Fluorescence of strain Ph6-*gfp* in culture solution (×100). Bacterial cells with green fluorescence were displayed as bright green rods on a glass microscope slide

Endophytes can not only abundantly reside in the internal tissues of plants but also have rich variety and different functions (Newman and Reynolds, 2005; Moore et al., 2006; Macek et al., 2008; Ryan et al., 2008). Until now, several endophytic

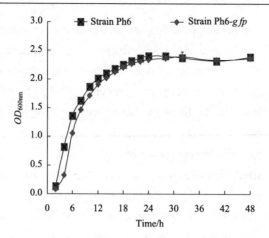

Figure 10-8　Growth curves of strains Ph6 and Ph6-*gfp*. Error bars are standard deviations (n = 3)

bacterial species were isolated from plants that grew well in the presence of naphthalene, catechol, and phenol, and it was discovered that these strains could utilize the pollutants as their sole carbon source and energy source for growth (Ho et al., 2009; Peng et al., 2013). A phenanthrene-degrading endophytic bacterium strain Ph6 was isolated from the interiors of PAH-contaminated clover plants, indicating that PAH-degrading endophytes can reside in the internal tissues of PAH-contaminated plants and may have specialized functions in plants (Weyens et al., 2009). Moreover, our result will also potentially increase the resources of PAH-degrading bacterial species and enrich the pool of PAH-degrading genes (Ryan et al., 2008; Peng et al., 2015).

10.2.2　Biodegradation of phenanthrene by Ph6-*gfp* in culture solution

Biodegradation of phenanthrene by Ph6-*gfp* was conducted in shake-flask culture. The growth and phenanthrene-degrading ability of strain Ph6-*gfp* were tested (Sun et al., 2014a). Strain *Pseudomonas* sp. Ph6-*gfp* was cultivated in MSM, and was adjusted to an approximate suspension of $2.5 \times 10^8 \sim 5.2 \times 10^8$ colony forming units (CFU/mL). 20 mL of fresh PMM media (Containing yeast extract and phenanthrene) was inoculated with the above suspension (1 mL) and was shaken at 30℃ and 150 r/min in a rotary shaker. The population size of strain Ph6-*gfp* was measured with plate counts. Phenanthrene residue in the flask was measured using HPLC.

Strain Ph6-*gfp* had phenanthrene degradation efficiency similar to that of the wild strain Ph6, which could degrade 85% of phenanthrene in a culture solution within

15 days. The degradation kinetics of phenanthrene and growth curves of strain Ph6-*gfp* were studied in the same culture solution as the wild strain (Figure 10-9). The results indicate that strain Ph6-*gfp* grew well and degraded phenanthrene effectively in 15 days. Cell counts of Ph6-*gfp* first rose and then fell after 15 days. The initial and maximum values (At day 5) of the cell counts were 8.60 and 8.72 (Lg CFU/mL), respectively. The concentration of phenanthrene in the culture solution with Ph6-*gfp* decreased consistently with time, and 81.1% of the phenanthrene was dissipated in 15 days. In the control solution (Ph6-*gfp*-free), only 21.6% of the phenanthrene disappeared due to abiotic dissipation in 15 days. This result indicates that Ph6-*gfp* was effective in phenanthrene degradation in culture medium. The data above show that the GFP gene did not disrupt any key traits required for survival or phenanthrene degradation of strain Ph6 and that Ph6-*gfp* could, in principle, be used to degrade phenanthrene in inner plant tissues.

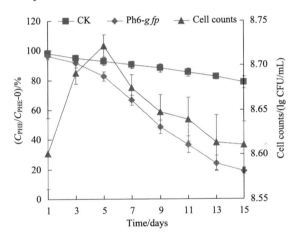

Figure 10-9 The degradation kinetics of phenanthrene and growth curves of strain Ph6-*gfp* in culture solution; C_{PHE} represents the residual concentration of phenanthrene in solution at a given time; C_{PHE-0} represents the initial concentration of phenanthrene in solution; Error bars are standard deviations

10.2.3 Colonization and distribution of Ph6-*gfp* in plants

Seedlings of ryegrass (*Lolium multiflorum* Lam.) were incubated in greenhouse until approximately 10 cm tall with relatively mature roots, and then they were transplanted into the Hoagland medium containing phenanthrene (1 mg/L) for 3 days, with roots submerged in the contaminated Hoagland medium. Plant roots were then washed with sterile water. Bacterial inoculum was prepared by resuspending pelleted cells in sterile water to obtain an inoculum density of approximately 9.5 (lg CFU/mL). Plant roots were thereafter immersed in a suspension of strain Ph6-*gfp* for 6 h. Then, the plant roots were re-rinsed four times with sterile water, dried on sterile filter paper, and replanted into a 300 mL brown glass bottle containing 250 mL of refresh Hoagland medium with 1 mg/L phenanthrene and roots submerged in the contaminated Hoagland medium. Plants with Ph6-*gfp*-free inoculation served as the control. Plant cultivation was carried out in environmental growth chambers set at 25/20℃ under a 12-hour light/dark regime. At 3 days, 6 days, 9 days, 12 days and 15 days exposure, plant samples were collected and prepared for PAH and bacteria analysis.

Colonization and distribution of Ph6-*gfp* in ryegrass were investigated. The quantities of strain Ph6-*gfp* in ryegrass roots, shoots, and in the Hoagland medium were determined by counting the CFU on plates. The localization and distribution of strain Ph6-*gfp* in ryegrass roots and shoots were monitored with fluorescence microscopy. Endophytic strain Ph6-*gfp* was an efficient colonizer of phenanthrene-contaminated ryegrass and could vertically transfer from roots to shoots. When ryegrass roots were soaked in a suspension of Ph6-*gfp* for 6 hours, Ph6-*gfp* first colonized the plant root surface and formed bacterial aggregates or biofilms. Strain Ph6-*gfp* then proceeded to enter into the plant roots (Figure 10-10D). As time progressed, Ph6-*gfp* cells were also visualized in the stems (Figure 10-10E) and leaves (Figure 10-10(F)). Plants with Ph6-*gfp*-free inoculation served as the control (Figure 10-10(A-C)). The quantification of Ph6-*gfp* within plant roots and shoots was performed by counting the CFU on plates (Table 10-5). The cell counts of Ph6-*gfp* decreased progressively from ryegrass roots to stems to leaves, and cell counts in the roots were significantly ($p<0.01$) higher than those in the shoots. The localization and distribution of strain Ph6-*gfp* within the plants changed with time. The cell counts of Ph6-*gfp* within the plants initially increased for the first 6 days and then gradually decreased from days 6~15. After inoculation for 15 days, the cell counts of Ph6-*gfp* were 5.51 lg CFU/g and 3.65 lg CFU/g in roots and shoots, respectively. In addition, strain Ph6-*gfp* could also be re-isolated and released into Hoagland solution, and cell counts were observed between 10^4 and 10^6 CFU/mL in solution. The colonization of Ph6-*gfp* provided the premise for the utilization of this endophytic strain to influence phenanthrene uptake by plants.

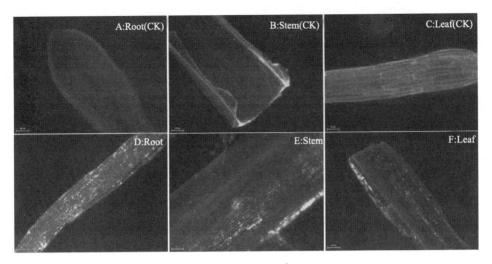

Figure 10-10 Visualization of inoculated endophyte Ph6-*gfp* within plant tissues. After inoculation with Ph6-*gfp* for 6 days, the colonization of Ph6-*gfp* was observed in plant roots, stems and leaves by fluorescence microscopy. Bacterial cells with green fluorescence were displayed as bright green rods inside roots (D), stems (E) and leaves (F). Plants with Ph6-*gfp*-free inoculation served as the control (A, B, C)

Table 10-5 Populations of Ph6-*gfp* in ryegrass roots, shoots, and culture solution as determined by counting the CFU on plates

Time/day	Root/(lg CFU/g)	Shoot/(lg CFU/g)	Culture solution/(lg CFU/mL)
3	5.66±0.04	4.07±0.05	4.78±0.07
6	5.79±0.03	4.60±0.04	6.02±0.08
9	5.70±0.02	4.08±0.13	5.96±0.07
12	5.60±0.04	3.94±0.08	5.02±0.03
15	5.51±0.05	3.65±0.10	4.90±0.02

Different plant colonization methods (seed soaking, SS; root soaking, RS; leaf painting, LP) of Ph6-*gfp* were also evaluated (Sun et al., 2015). A semi-closed system was performed in a greenhouse to compare the efficiencies of different inoculation methods (Figure 10-11). Colonization methods, including SS, RS, and LP, were used for inoculation with strain Ph6-*gfp*. SS was processed as follows. In brief, surface-sterilized plant seeds were immersed into a suspension with strain Ph6-*gfp* for 6 h, rinsed four times with sterile water, and incubated at 30℃ for germination. When seedlings were approximately 10 cm tall, strong individuals were transplanted into a brown glass pot containing 250 mL Hoagland medium (1 mg/L phenanthrene) for 15 days. RS and LP were processed as follows. Strong ryegrass seedlings (approximately 10 cm tall) were grown in Hoagland medium containing 1 mg/L phenanthrene (PHE) for 9 days. Plant roots were then soaked into a suspension with strain Ph6-*gfp* for 6 h, or plant leaves were painted with the suspension three times (20 min each time). Then, plant roots/leaves were rinsed four times with sterile water, dried with sterile filter paper, and cultivated in the original Hoagland medium for 6 days.

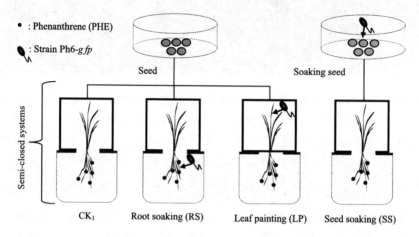

Figure 10-11 Diagram of the greenhouse experimental design in which a semi-closed system was used. Colonization methods were (1) seed soaking (SS), (2) root soaking (RS), and (3) leaf painting (LP). CK_1 indicates the control

We enumerated cells of strain Ph6-*gfp* in ryegrass by counting CFUs on plates (Table 10-6). The cell numbers of Ph6-*gfp* in various tissues of inoculated ryegrass showed efficient colonization of inoculated plants. Furthermore, Ph6-*gfp* most successfully colonized roots compared to seeds and shoots. Ph6-*gfp* colonized plant seeds via SS for 6 h (Figure 10-12A), and the populations observed in seeds were 4.7 lg CFU/g. Similarly, after RS for 6 h, Ph6-*gfp* first resided in large biofilms on root surfaces, and then actively and internally colonized ryegrass plant roots (Figure 10-12B). As time progressed, Ph6-*gfp* transferred vertically to the stems (Figure 10-12C) and leaves (Figure 10-12D). 6 days after colonization via RS, the quantification of Ph6-*gfp* within plant roots and shoots was 5.8 and 4.7 lg CFU/g, respectively. In addition, Ph6-*gfp* colonized target plant tissues from the phyllosphere to internal tissues and was also found inside stems after LP (Figure 10-12E, F). After 6 days via LP, the colonies of Ph6-*gfp* cells in the interior shoot tissues of inoculated plants were 5.5 lg CFU/g, but there was no fluorescence detected in plant roots. In contrast, fluorescence remained undetectable in non-inoculated plants.

Table 10-6   The cell numbers (log CFU/g, on a fresh weight basis) of endophytic *Pseudomonas* sp. Ph6-*gfp* in interior tissues of inoculated ryegrass seeds, roots, and shoots

Colonization methods	Seed/(lg CFU/g)	Root/(lg CFU/g)	Shoot/(lg CFU/g)
Seed soaking (SS)	4.7±0.2	ND	ND
Root soaking (RS)	—	5.8±0.1	4.7±0.2
Leaf painting (LP)	—	ND	5.5±0.1

Note: ND indicates that Ph6-*gfp* was undetectable in the interior tissues of plants. Errors bars ± are standard deviations.

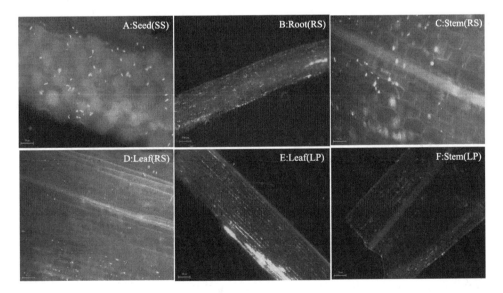

Figure 10-12   Endophytic *Pseudomonas* sp. Ph6-*gfp* cells in seeds, roots, stems, and leaves of inoculated plants were exposed to 1 mg/L of PHE via three colonization methods. Ph6-*gfp* colonization via SS was observed in plant seeds using fluorescence microscopy (A). After inoculation with Ph6-*gfp* via RS for 6 days, colonization was observed in plant roots, stems, and leaves (B~D). After inoculation with Ph6-*gfp* via LP, colonization was observed in plant leaves and stems (E, F)

10.2.4 Performances of Ph6-*gfp* mediate the uptake of phenanthrene by plants

Strain Ph6-*gfp* could significantly reduce the risks associated with plant phenanthrene-contamination based on observations of decreased concentration and accumulation of phenanthrene in plant bodies. Furthermore, the translocation of phenanthrene from roots to shoots decreased in endophyte-colonized plants.

Strain Ph6-*gfp* showed a natural capacity to cope with phenanthrene *in vitro* as well as within plants (Figures 10-9 and 10-13). The concentrations of phenanthrene in ryegrass roots and shoots with and without Ph6-*gfp* inoculation (RS) are shown in Figure 10-13. After Ph6-*gfp* inoculation for 9~15 days, the concentrations of phenanthrene in endophyte-colonized roots and shoots were 18.8~89.8 mg/kg and 1.25~3.82 mg/kg, respectively, whereas the concentrations in roots and shoots colonized by the Ph6-*gfp*-free inoculation were 24.5~110 mg/kg and 5.49~6.89 mg/kg, respectively. The concentrations of phenanthrene in endophyte-colonized roots and shoots decreased in 9~15 days. Compared with the Ph6-*gfp*-free treatment, the respective concentrations of phenanthrene in endophyte-colonized roots and shoots were 18.5%~23.3% and 30.4%~81.1% lower, respectively. With the inoculation with Ph6-*gfp*, a more significant decrease in phenanthrene concentration was observed in the shoots than in the roots. These results indicate that Ph6-*gfp* aided phenanthrene degradation in contaminated ryegrass and significantly reduced plant phenanthrene contamination. In addition, the concentrations of phenanthrene in roots were always one to two magnitudes larger than those in shoots, irrespective of the inoculation with Ph6-*gfp*.

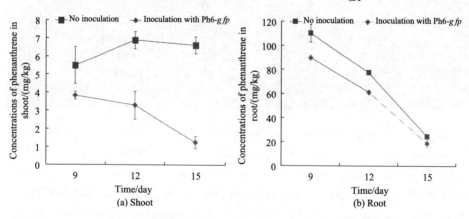

Figure 10-13 The concentrations of phenanthrene in ryegrass shoots (a) and roots (b) as a function of time. Error bars are standard deviations

The inoculation of ryegrass with strain Ph6-*gfp* can also reduce the accumulated amounts (*A*, μg/bottle) of phenanthrene in plants. *A* was estimated according to the following equation

$$A = C \times M$$

where *C* (mg/kg) is the concentration of phenanthrene in the root or shoot, and *M* (g/bottle, on a dry weight basis) is the biomass of ryegrass root or shoot in each bottle. Higher *A* values indicate more phenanthrene present in the plants and higher risk of plant contamination. The calculated *A* values are given in Table 10-7. The accumulation of phenanthrene in ryegrass shoots was much smaller than in roots, irrespective of Ph6-*gfp* inoculation (RS); i.e., the root was the dominant sink of the phenanthrene present in the plant bodies, although shoots were also extensively contaminated by phenanthrene. After 9~15 days, the accumulated amounts of phenanthrene in endophyte-colonized roots and shoots were 0.87~2.30 μg/bottle and 0.17~0.31 μg/bottle, respectively, corresponding to values 10.3%~17.6% and 30.8%~66.5% lower than those in Ph6-*gfp*-free plant roots and shoots. This result indicates that the inoculation of ryegrass with Ph6-*gfp* reduced phenanthrene accumulation in plant bodies. Moreover, similar to the trend of plant phenanthrene concentrations, inoculation with Ph6-*gfp* produced a decrease in phenanthrene accumulation that was more significant in the ryegrass shoots than in the roots. In addition, compared with those at day 12, the *A* values in roots and shoots were smaller at day 15, irrespective of the inoculation with Ph6-*gfp*, suggesting the obvious metabolism of phenanthrene in plants.

Table 10-7 The accumulation amounts of phenanthrene in ryegrass roots and shoots

Time/day	No inoculation		Inoculation of Ph6-*gfp*	
	Root/(μg/pot)	Shoot/(μg/pot)	Root/(μg/pot)	Shoot/(μg/pot)
9	3.64±0.24	0.39±0.07	2.30±0.08	0.27±0.02
12	2.85±0.01	0.55±0.04	2.41±0.07	0.31±0.07
15	0.97±0.02	0.50±0.03	0.87±0.15	0.17±0.04

The translocation factor (TF, unitless) of phenanthrene in ryegrass was estimated as

$$TF = SCF / RCF$$

where RCF and SCF are the root and shoot concentration factors, respectively. RCF and SCF were obtained according to the following equations

$$SCF = C_{shoot} / C_w$$
$$RCF = C_{root} / C_w$$

Therefore, by combining above Equations, TF could be further estimated as

$$TF = C_{shoot} / C_{root}$$

where C_{shoot}, C_{root}, and C_w are the phenanthrene concentrations in plant shoots, plant roots, and Hoagland solution, respectively. A larger TF value infers a more significant translocation of phenanthrene from root to shoot. As shown in Figure 10-14, the TF values increased from day 9 to day 15, irrespective of Ph6-*gfp* inoculation (RS). The TF values of phenanthrene in Ph6-*gfp*-free ryegrass in day 9~15 were 0.05~0.27, which were 25.0%~286% greater than that in Ph6-*gfp*-inoculated plants. These results indicate that strain Ph6-*gfp* efficiently impeded phenanthrene translocation from the roots to the shoots of ryegrass.

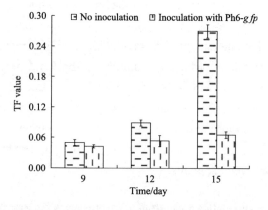

Figure 10-14 The translocation factor (TF) of phenanthrene in ryegrass as a function of uptake time. Error bars are standard deviations

Pseudomonas sp. can be beneficial in the protection of plants from soils contaminated by organic pollutants (Ryan et al., 2008; Germaine et al., 2006,2009). The poplar endophyte *Pseudomonas putida* VM1441 was an efficient colonizer of interior plant root tissues and allowed the host plant to circumvent the phytotoxic effects of naphthalene (Germaine et al., 2009). Furthermore, with the colonization of the endophytic bacterium *Pseudomonas putida* 1450, the pea plant *Pisum sativum* displayed no 2,4-dichlorophenoxyacetic acid (2,4-D) accumulation in plant aerial tissues (Germaine et al., 2006). However, few previous investigations have tracked the visualization to elucidate the performance of PAH-degrading functional endophytes in plants. We demonstrated that the PAH-degrading endophytic bacterium *Pseudomonas* sp. Ph6 tagged with the GFP gene can re-colonize PAH-contaminated plants and be transferred from roots to stems and leaves. Furthermore, Ph6-*gfp* has a natural capacity to cope with phenanthrene *in planta* (Figure 10-13), which provides the possibility of utilizing endophyte Ph6-*gfp* to degrade PAHs in plants.

Large quantities of PAH-degrading endophytes colonize plant interiors and play a significant role by promoting the degradation of PAHs in plants. The endophytic populations observed in spruce were between 2.0 lg CFU/g and 7.0 lg CFU/g FW^{37}, and the colonies of *Pseudomonas* sp. cells in the interior root and shoot tissues of inoculated poplar trees were between 2.0 lg CFU/gFW and 5.0 lg CFU/g FW (Germaine et al., 2004). In our study, after the ryegrass roots were soaked with the suspension of strain Ph6-*gfp*, the highest densities of strain Ph6-*gfp* were observed in the roots (5.51~5.79 lg CFU/g FW), and densities decayed progressively from the stems to the leaves. This result was similar to that observed for *Herbaspirillum seropedicae* sp. in rice seedlings, even though the concentration of *H. seropedicae* sp. cells was much higher (6 lg CFU/g FW) (James et al., 2002). This indicates that PAH-degrading endophytes can readily adapt to inner plant circumstances and aid in the elimination of the risks involved with plant PAH contamination.

The possible mechanisms involved in the endophyte-enhanced degradation of PAHs in plants include the following: ①Endophytes can be released into soil and promote PAH degradation in the environment and thereby reduce the uptake of PAHs by plant roots (Sun et al., 2014b); ②Endophytes not only degrade PAHs *in vitro* but can have the same effect *in planta* (Sun et al., 2014b); ③Endophytic inoculants can transfer their PAH-degradative genes to other endophytes in plants, thus increasing the overall degradation potential of endogenous endophytic communities (Taghavi et al., 2005; Wang et al., 2007); ④Endophytes may influence the activities of plant enzymes, such as oxidase, reductase, esterase, and dehalogenase, participating in the transformation of PAH contaminants (Harish et al., 2009; Weyens et al., 2009);⑤Endophytes may promote PAH-metabolizing gene expression, such as alkane monooxygenase (*alk*B) and naphthalene dioxygenase (*ndo*B) (Siciliano et al., 2001).

10.3 Utilizing endophytic bacterium *Staphylococcus* sp. BJ06 to reduce plant pyrene contamination

In this section, a pyrene-degrading endophytic bacterium, *Staphylococcus* sp. BJ06, was isolated from *Alopecurus aequalis* grown in PAH-contaminated sites. After root inoculation with strain BJ06, greenhouse container experiments were conducted to evaluate the performance of BJ06-mediated uptake of phenanthrene by ryegrass (*Lolium multiflorum* Lam) in soil.

10.3.1 Isolation and identification of *Staphylococcus* sp. BJ06

Several plant species (*Alopecurus aequalis*, *Trifolium pratense* L., and *Conyza canadensis*) growing on PAH-contaminated sites were collected for the isolation of pyrene-degrading endophytic bacteria using enrichment culture. The isolation protocol and identification was detailed by Sun et al. (2014b). The physiological and biochemical characteristics of strain BJ06 were obtained from a bacterial identification manual (George et al., 2004). A phylogenetic tree was constructed from the evolutionary distances using the neighbor-joining method. The tree topologies were evaluated by performing a bootstrap analysis of 1000 datasets with the MEGA 5.05 package.

A pyrene-degrading endophytic bacterium, BJ06 (Figure 10-15), was isolated from *Alopecurus aequalis*. Strain BJ06 grew well on plates containing naphthalene (200 mg/L), phenanthrene (100 mg/L), or benzo[a]pyrene (10 mg/L). This strain is a globular, gram-positive bacterium with flagella. The indole test and methyl red test were negative, and the Voges-Prokauer test was positive. Furthermore, strain BJ06 was not positive for nitrate reductase, phenylalanine dehydrogenase, catalase, or citrate. This strain did not hydrolyze starch or gelatin and did not grow on plates with medium containing glucose, fructose, or sucrose. A blast of the strain BJ06 16S rDNA sequence with sequences in Genbank showed that strain BJ06 is more than 99% identical to strains of *Staphylococcus* sp. The GenBank accession number of the 16S rDNA sequence of strain BJ06 is KC236189. A phylogenetic tree that includes strain BJ06 and related species is presented in Figure 10-16. On the basis of its morphology, physiology, and 16S rDNA sequence, strain BJ06 was identified as a strain of *Staphylococcus* sp..

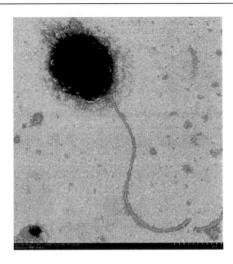

Figure 10-15 Transmission electron micrograph of strain BJ06 isolated from *Alopecurus aequalis*

(×4.0K Zoom-1 HC-1 80 kV)

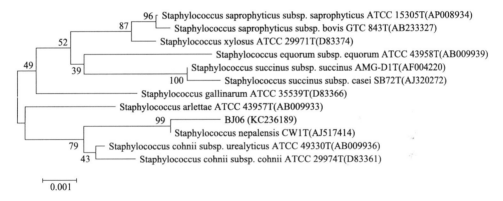

Figure 10-16 Phylogenetic analysis of the isolate BJ06 and related species using the neighbor-joining method. Bootstrap values (%) are indicated at the nodes in a bootstrap analysis of 1000 replicates. The scale bar indicates 0.001 changes per nucleotide. The GenBank accession number for each microorganism used in the analysis is shown in parentheses after the species name

10.3.2 Biodegradation of pyrene by BJ06 in culture solution

Strain BJ06 was cultivated in LB medium. Cells were collected by centrifugation at 8000 r/min for 10 min, washed twice with MSM, and resuspended at $2.5 \times 10^8 \sim 5.2 \times 10^8$ colony forming units (CFU/mL). The cells were inoculated at the 5% (v/v) level into a 50 mL flask containing 20 mL PMM and incubated at 30℃ and 150 r/min in a rotary shaker. The growth and ability to degrade pyrene of strain BJ06 were determined after cell inoculation with PMM. The effects of incubation pH (4.0~9.0), temperature (15~45℃), NaCl concentration (0~30 g/L), liquid volume (10~70 static), inoculum size (1%~15%), and initial pyrene concentration (10~70 mg/L) on pyrene biodegradation by strain BJ06 were studied. Media containing inactivated BJ06 served as a control (Sun et al., 2014b).

The dynamics of pyrene degradation were studied in liquid PMM inoculated with strain BJ06 (Figure 10-17). Strain BJ06 grew efficiently with pyrene as the sole carbon source. The direct counting of CFU on PMM plates showed an initial decrease in the cell count on day 1; thereafter, the numbers increased gradually as pyrene was utilized on days 1~9. After 9 days of incubation, the cell count decreased concurrently with the decrease in pyrene concentration. The initial and maximum values of the cell counts were 7.7 and 8.1 (lg CFU/mL), respectively. Strain BJ06 degraded 56% of the pyrene in 15 days.

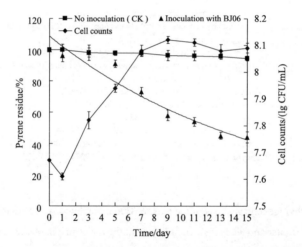

Figure 10-17 The degradation dynamics of pyrene and growth curve of strain BJ06 with pyrene as its sole carbon source. Error bars are ± standard deviation

Optimization of the pyrene-degrading conditions could enhance the rate of pyrene degradation of strain BJ06. The degradation rate of pyrene (50 mg/L) by strain BJ06 was 60.6% in 15 days at the optimal pH of 6.0, whereas the degradation rate was decreased at an initial pH above 8.0 or below 6.0 (Figure 10-18(a)). Strain BJ06 degraded 56.8% of pyrene at the optimal temperature of 30℃, but the degradation rates at 20℃ and 40℃ were approximately half as fast as at 30℃ (Figure 10-18(b)).

The optimum NaCl concentration for pyrene biodegradation by strain BJ06 (Figure 10-18) was 0~15 g/L. Under these conditions, strain BJ06 grew and degraded pyrene effectively (Figure 10-18(c)). The results of the oxygen supply test showed that the pyrene degradation rate and cell growth rate of BJ06 were negatively correlated with increasing liquid volume (Figure 10-18(d)). Therefore, strain BJ06 grew aerobically.

The inoculum density and initial pyrene concentration significantly affected the degradation rate of pyrene by strain BJ06. The degradation of 50 mg/L pyrene after 15 days of incubation at 1%, 3%, 5%, 7%, 9%, 12% and 15% initial inoculum density ranged from 12.7% to 65.9% (Figure 10-18(e)). Clearly, the pyrene degradation rate varied in direct proportion to the inoculum density of strain BJ06. Increases in the initial pyrene concentration decreased the pyrene degradation rate by strain BJ06 (Figure 10-18(f)).

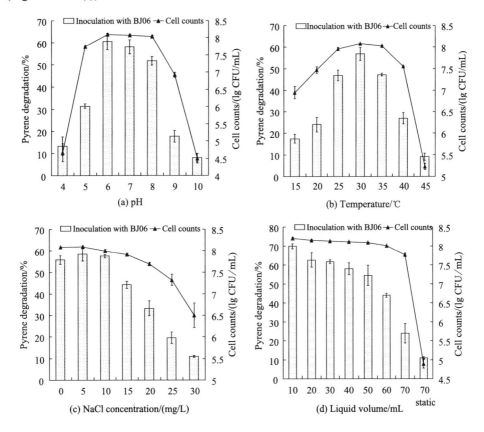

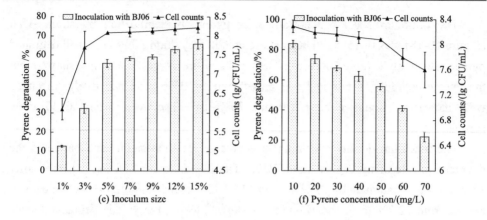

Figure 10-18 Effects of different environmental conditions on pyrene degradation and growth curves of strain BJ06. Error bars are ± standard deviation

The metabolic pathways of pyrene degradation in *Staphylococcus* sp. BJ06 are currently unclear. Phthalic acid, salicylic acid, and catechol were reported to be the typical metabolic products of pyrene degradation in bacterial strains (Rehmann et al., 1998). Some of the metabolic products from a pyrene-degrading bacterium (*Mycobacterium flavescens*) that was isolated from PAH-polluted sediments have been identified. These products include *cis*-4, 5-pyrene dihydrodiol, 4-5-phenanthrenedioic acid, 4-phenanthroic acid, and phthalate acid (Dean-Ross and Cerniglia 1996). Nevertheless, there have been no reports that identify the metabolic products of pyrene in strains of *Staphylococcus* sp. In this study, the metabolites of pyrene in *Staphylococcus* sp. BJ06 were extracted and analyzed. The results indicated that two of the metabolites of pyrene in *Staphylococcus* sp. BJ06 were phthalate acid and catechol. These two metabolites were also detected in the metabolic pathway of phenanthrene in *Staphylococcus* sp. strain PN/Y (Somnath et al., 2007). However, there is still much work that needs to be done to identify the products of pyrene degradation and to clarify the metabolic pathways for strain BJ06.

10.3.3 Reducing plant pyrene contamination using strain BJ06

Yellow-brown soil was collected from the soil surface (0~20 cm) in Nanjing, China. The soil had a pH of 6.02 and was composed of 14.3 g/kg soil organic carbon, 24.7% clay and 67.9% sand. Soil samples were pyrene-spiked, aged and potted for ryegrass vegetation (Sun et al., 2014b). The container experiment consisted of two levels of pyrene contamination (0 mg/kg or 100 mg/kg) and six treatments (uncontaminated soil + ryegrass (UR), uncontaminated soil + ryegrass + bacteria (URB), contaminated soil (CK), contaminated soil + ryegrass (CR), contaminated soil + bacteria (CB), and contaminated soil + ryegrass + bacteria (CRB)). A bacterial suspension in MSM (5 mL/pot) was inoculated to unplanted (CB) and planted (URB and CRB) soil surfaces 10 days after seedling emergence, and a sterilized bacterial suspension in MSM was applied (UR, CK, and CR) as a control. The ryegrass seedlings were grown for 15 days in a greenhouse. Soil and plant samples were collected and prepared for PAH analysis.

Strain BJ06 colonized ryegrass tissues and increased ryegrass growth. After 15 days of incubation with pyrene, ryegrass growth did not exhibit significant stress or toxic effects (Table 10-8). In non-polluted soils, ryegrass height increased from 14.17 m to 15.07 cm and root length decreased from 6.22 cm to 4.79 cm after colonization by BJ06. In polluted soils, ryegrass height increased from 13.00 cm to 14.14 cm and root length decreased from 6.00 cm to 5.71 cm with BJ06 colonization. Therefore, significant adverse effects of strain BJ06 on plant height and root length ($p<0.05$) were observed. Compared with the endophyte-free treatment, the ryegrass fresh weight and dry weight were both significantly higher in the endophyte-inoculated treatment group.

Table 10-8 Ryegrass biomass and cell counts of BJ06 in ryegrass after plants were colonized by BJ06 and inoculated in yellow-brown soil for 15 days

Treatments	Root (colonization)			
	Length /cm	Fresh weight /(mg/pot)	Dry weight /(mg/pot)	Cell counts /($\times 10^3$ CFU/g of fresh plant)
UR	6.22±0.41	92.61±6.54	21.00±2.32	ND
URB	4.79±0.51	107.06±4.36	26.44±2.24	83.50±10.60
CR	6.00±0.98	114.72±10.31	25.72±2.11	ND
CRB	5.70±0.41	125.39±9.56	33.89±6.26	146.00±8.50
Treatments	Shoot (colonization)			
	Length /cm	Fresh weight /(mg/pot)	Dry weight /(mg/pot)	Cell counts /($\times 10^3$ CFU/g of fresh plant)
UR	14.17±0.64	293.72±17.41	21.17±1.83	ND
URB	15.07±0.71	338.56±31.21	22.83±2.52	14.50±0.70
CR	13.00±0.47	289.00±31.14	19.78±1.34	ND
CRB	14.14±1.10	293.78±9.42	20.44±0.82	51.50±7.80

Note: UR-planted ryegrass in uncontaminated soil, URB-planted ryegrass and inoculated with strain BJ06 in uncontaminated soil, CR-planted ryegrass in contaminated soil, CRB-planted ryegrass and inoculated with strain BJ06 in contaminated soil. Error bars are ± standard deviation. ND means undetectable.

After 15 days inoculation in uncontaminated soil, the cell count of BJ06 was 83.5×10^3 CFU/g fresh weight in roots and 14.5×10^3 CFU/g in shoots. In contaminated soil, the cell counts were 146×10^3 CFU/g and 51.1×10^3 CFU/g in roots and shoots, respectively (Table 10-8). The cell counts of strain BJ06 in ryegrass tissues in contaminated soil were significantly ($p<0.01$) higher than those in uncontaminated soil, and these counts increased by 74.9% and 252% in roots and shoots, respectively.

The uptake of pyrene from soil by ryegrass roots and shoots after 15 days is shown in Table 10-9. In contaminated soil, more pyrene accumulated in the roots than in the shoots. The concentration of pyrene in ryegrass roots and shoots inoculated with BJ06 was 54 mg/kg and 4 mg/kg, respectively. However, without BJ06 inoculation, the pyrene concentrations were 79 mg/kg and 6 mg/kg in roots and shoots ($p<0.05$), respectively. The pyrene accumulation in roots decreased from 2.03×10^{-3} mg/pot to 1.85×10^{-3} mg/pot with BJ06 inoculation, and in shoots, the accumulation decreased from 0.13×10^{-3} mg/pot to 0.07×10^{-3} mg/pot. The plant concentration factor (PCF) describes the capability of plants to accumulate contaminants from direct contact with the soil environment. In this study, PCF was calculated as the ratio of pyrene concentration in plant roots and shoots (C_p) to the pyrene concentration in the soil (C_s) (PCF=C_p/C_s). The plant translocation factor (TF) is defined as the ratio of pyrene concentration in the shoot concentration factor (SCF) to the pyrene concentration in the root concentration factor (RCF) (TF= SCF/RCF). A larger TF value indicates that more pyrene is translocated by ryegrass from roots to shoots. The PCF and TF of pyrene in ryegrass with BJ06 inoculation (0.05~0.77, 0.06) were significantly lower than in the control treatment without inoculation (0.08~1.03, 0.08) after 15 days.

Table 10-9 Concentration of pyrene in ryegrass planted in yellow-brown soil with 100 mg/kg of pyrene for 15 days

Treatments	The concentration of pyrene (C_p)		Plant concentration factors (PCF)		Accumulation		Translocation factor (TF)
	Root /(mg/kg)	Shoot /(mg/kg)	Root /(C_p/C_s)	Shoot /(C_p/C_s)	Root /(mg/pot)	Shoot /(mg/pot)	
CK	ND	ND	—	—	—	—	—
CR	78.92±6.31	6.31±0.44	1.03	0.08	2.03×10^{-3}	0.13×10^{-3}	0.08
CB	ND	ND	—	—	—	—	—
CRB	54.45±4.32	3.52±0.19	0.77	0.05	1.85×10^{-3}	0.07×10^{-3}	0.06

Note: CK-contaminated soil, CR-planted ryegrass in contaminated soil, CB-inoculated with strain BJ06 in contaminated soil, CRB-planted ryegrass and inoculated with strain BJ06 in contaminated soil. Error bars are ± standard deviation. ND means undetectable.

In addition, the pyrene removal in planted soil (CR) or endophyte-inoculated soil (CB) was significantly increased compared to contaminated soil without ryegrass or endophytic bacterium inoculation (CK). After 15 days, the rate of pyrene removal in soil contaminated with 100 mg/kg of pyrene was 7% in CK, while in CR and CB, the rate of removal was 23% and 13%, respectively. Furthermore, in planted soil with endophyte inoculation (CRB), the rate of pyrene removal was 29% (Figure 10-19). In brief, ryegrass or endophyte inoculation significantly enhanced the rate of pyrene removal in contaminated soil, and the enhancement was greater in the endophyte-inoculated and ryegrass planted soil ($p<0.05$).

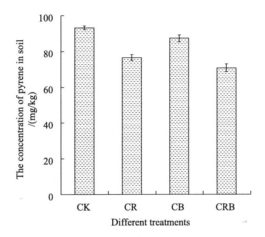

Figure 10-19 Pyrene concentrations in unplanted and planted soil that was not inoculated or inoculated with endophytic pyrene-degrading bacterial strain BJ06. Yellow-brown soil with 100 mg/kg of pyrene added was used. Error bars are ± standard deviation

In summary, a pyrene-degrading endophytic bacterium, *Staphylococcus* sp. BJ06, was isolated from *Alopecurus aequalis*. This is the first report of a pyrene-degrading bacterium in the genus of *Staphylococcus* sp. BJ06 grew and degraded pyrene efficiently under environmental conditions. The bacterium significantly promoted ryegrass growth and pyrene removal from inner plant and contaminated soil in container experiments. We provide new perspectives on the regulation and control of plant uptake of organic contaminants with endophytic bacteria. The results of this study will be valuable to risk assessments of plant PAH contamination.

References

Aajoud A, Raveton M, Aouadi H, et al. 2006. Uptake and xylem transport of fipronil in sunflower. J. Agric. Food Chem., 54: 5055~5060.

Alexander M. 2000. Aging, bioavailability, and overestimation of risk from environmental pollutants. Environ. Sci. Technol., 34:4259~4265.

Alkio M, Tabuchil T M, Wang X, et al. 2005. Stress responses to polycyclic aromatic hydrocarbons in Arabidopsis include growth inhibition and hypersensitive response-like symptoms. J. Exp. Bot., 56: 2983~2994.

Alkorta I, Garbisu C. 2001. Phytoremediation of organic contaminants in soils. Bioresour. Technol., 79: 273~276.

An C J, Huang G H, Wei J, et al. 2011. Effect of short-chain organic acids on the enhanced desorption of phenanthrene by rhamnolipid biosurfactant in soil-water environment. Water Res., 45: 5501~5510.

Araújo W L, Marcon J, Maccheroni W, et al. 2002. Diversity of endophytic bacterial populations and their interaction with *Xylella fastidiosa* in citrus plants. Appl. Environ. Microbiol., 68: 4906~4914.

Baborová P, Möder M, Baldrian P, et al. 2006. Purification of a new manganese peroxidase of the white-rot fungus *Irpex lacteus*, and degradation of polycyclic aromatic hydrocarbons by the enzyme. Res. Microbiol., 157: 248~253.

Bacci E, Cerejeira M J, Gaggi C, et al. 1990. Bioconcentration of organic chemical vapours in plant leaves: the Azalea model. Chemosphere, 21: 525~535.

Bais H P, Weir T L, Perry L G, et al. 2006. The role of root exudates in rhizosphere interactions with plants and other organisms. Ann. Rev. Plant Biol., 57: 233~266.

Barraclough D, Kearney T, Croxford A. 2005. Bound residues: environmental solution or future problem. Environ. Pollut., 133: 85~90.

Barriuso E, Andrades M S, Benoit P, et al. 2011. Pesticide desorption from soils facilitated by dissolved organic matter coming from composts: experimental data and modelling approach. Biogeochem., 106: 117~133.

Binet P, Portal J M, Leyval C. 2000. Dissipation of 3ring~6-ring polycyclic aromatic hydrocarbons in the rhizosphere of ryegrass. Soil. Biol. Biochem., 32: 2011~2017.

Bogan B L, Lamar R T. 1995. One-electron oxidation in the degradation of creosote polycyclic aromatic hydrocarbons by *Phanerochaete chrysosporium*. Appl. Environ. Microbiol., 61: 2631~2635.

Brady C A L, Gill R A, Lynch P T. 2003. Preliminary evidence for the metabolism of benzo[a]pyrene by *plantago lanceolata*. Environ. Geochem Health., 25: 131~137.

Briggs G G, Bromilow R H, Evans A A. 1982. Relationship between lipophilicity and root uptake and translocation of non-ionized chemicals by barley. Pestic. Sci., 13: 495~504.

Briggs G G, Bromilow R H, Evans A A, et al. 1983. Relationships between lipophilicity and the distribution of non-ionized chemicals in barley shoots following uptake by the roots. Pestic. Sci., 14: 492~500.

Brion D, Pelletier E. 2005. Modeling PAHs adsorption and sequestration in freshwater and marine sediments. Chemosphere, 61:867~876.

Bromilow R H, Chamberlain K. 1995. Principles Governing Uptake and Transport of Chemicals// Trapp S, McFarlane J C. in Plant Contamination: Modelling and Simulation of Organic Chemical Processes. Lewis: Boca Raton.

Brooksd D S, Gonzalez C F, Appel DN, et al. 1994. Evaluation of endophytic bacteria as potential biological control agents for oak wilt. Biol. Control., 4: 373~381.

Burken J G, Schnoor J L. 1998. Predictive relationships for uptake of organic contaminants by hybrid poplar trees. Environ. Sci. Technol., 32: 3379~3385.

Burkert B, Robson A. 1994. ^{65}Zn uptake in subterranean clover (*Trifolium subterraneum* L.) by three vesicular-arbuscular mycorrhizal fungi in root-free sandy soil. Soil. Biol. Biochem., 26: 1117~1124.

Cajthaml T, Moder M, Kacer P, et al. 2002. Study of fungal degradation products of polycyclic aromatic hydrocarbons using gas chromatography with ion trap mass spectrometry detection. J. Chromatogr. A., 974: 213~222.

Carmichael L M, Christman R F, Pfaender F K. 1997. Desorption and mineralization kinetics of phenanthrene and chrysene in contaminated soils. Environ. Sci. Technol., 31: 126~132.

Carrier P, Baryla A, Havaux M. 2003 Cadmium distribution and microlocalization in oilseed rape (*Brassica napus*) after long-term growth on cadmium-contaminated soil. Planta., 216: 939~950.

Celis R, Barriuso E, Houot S. 1998. Effect of liquid sewadge addition on atrazine sorption and desorption by soil. Chemosphere, 37: 1091~1107.

Chelius M, Triplett E. 2001. The Diversity of Archaea and Bacteria in Association with the Roots of *Zea mays* L. Microbial Ecol., 41: 252~263.

Chen B D, Li X L, Tao H Q, et al. 2003. The role of arbuscular mycorrhiza in zinc uptake by red clover growing in a calcareous soil spiked with various quantities of zinc. Chemosphere, 50: 839~846.

Chen W M, Tang Y Q, Mori K, et al. 2012. Distribution of culturable endophytic bacteria in aquatic plants and their potential for bioremediation in polluted waters. Aquat. Biol., 15: 99~110.

Cheng Z X, Ling W T, Gao Y Z, et al. 2008. Impacts of arbuscular mycorrhizae on plant uptake and phytoremediation of pyrene in soils. Plant Nutr. Fertil. Sci., 14: 1178~1185 (in Chinese).

Cheng K Y, Wong J W C. 2006. Combined effect of nonionic surfacant Tween 80 and DOM on the behaviors of PAHs in soil-water system. Chemosphere, 62: 1907~1916.

Chilom G, Bestetti G, Sello G, et al. 2004. Formation of bound residues by naphthalene and *cis*-naphthalene-1,2-dihydrodiol. Chemosphere, 56 : 853~860.

Chiou C T, Malcolm R L, Brinton T I, et al. 1986. Water solubility enhancement of some organic pollutants and pesticides by dissolved humic and fulvic acids. Environ. Sci. Technol., 20: 502~508.

Chiou C T, McGroddy S E, Kile D E. 1998. Partition characteristics of polycyclic aromatic hydrocarbons on soils and sediments. Environ. Sci. Technol., 32: 264~268.

Chiou C T, Peters L J, Freed V H. 1979. A physical concept of soil–water equilibria for nonionic organic compounds. Science, 206: 831.

Chiou C T, Porter P E, Schmedding D W. 1983. Partition equilibria of nonionic organic compounds between soil organic matter and water. Environ. Sci. Technol., 17: 227~231.

Chiou C T, Sheng G, Manes M. 2001. A partition-limited model for the plant uptake of organic contaminants from soil and water. Environ. Sci. Technol., 235: 1437~1444.

Christensen B T. 2001. Physical fractionation of soil and structural and functional complexity in organic matter turnover. Europ. J. Soil. Sci., 52: 345~353.

Chroma L, Macek T, Demnerova K, et al. 2002a. Decolorization of RBBR by plant cells and correlation with the transformation of PCBs. Chemosphere, 49: 739~748.

Chroma L, Mackova M, Kucerova P, et al. 2002b. Enzymes in plant metabolism of PCBs and PAHs. Acta. Biotechnol, 22: 35~41.

Chung N, Alexander M. 1998. Differences in sequestration and bioavailability of organic compounds aged in dissimilar soils. Environ. Sci. Technol., 32: 855~860.

Chung N, Alexander M. 2002. Effect of soil properties on bioavailability and extractability of phenanthrene and atrazine sequestered in soil. Chemosphere, 48: 108~115.

Collins C D, Fryer M, Grosso A. 2006. Plant uptake of non-ionic organic chemicals. Environ. Sci. Technol., 40: 45~52.

Compant S, Clement C, Sessitsch A. 2010. Plant growth-promoting bacteria in the rhizo-and endosphere of plants: Their role, colonization, mechanisms involved and prospects for utilization. Soil. Biol. Biochem., 42: 669~678.

Corgié S C, Beguiristain T, Leyval C. 2004. Spatial distribution of bacterial communities and phenanthrene degradation in the rhizosphere of *Lolium perenne* L. Appl. Environ. Microbiol, 70: 3552~3557.

Corgié S C, Joner E J, Leyval C. 2003. Rhizospheric degradation of phenanthrene is a function of proximity to roots. Plant Soil., 257:143~150.

Costa R, Götz M, Mrotzek N, et al. 2006. Effects of site and plant species on rhizosphere community structure as revealed by molecular analysis of microbial guilds. FEMS Microbiol Ecol., 56: 236~249.

Cousins I T, Mackay D. 2001. Strategies for including vegetation compartments in multimedia models. Chemosphere, 44: 643~654.

Cox L, Celis R, Hermosin M C, et al. 2000. Effect of organic amendments on herbicide sorption as related to the nature of the dissolved organic matter. Environ. Sci. Technol., 34 : 4600~4605.

Criquet S, Joner E, Leglize P, et al. 2000. Anthracene and mycorrhiza affect the activity of oxidoreductases in the roots and the rhizosphere of lucerne (*Medicago sativa* L.). Biotechnol Lett., 22: 1733~1737.

Cuypers C, Pancras T, Grotenhuis T, et al. 2002. The estimation of PAH bioavailability in contaminated sediments using hydroxypropyl-β-cyclodextrin and triton X-100 extraction techniques. Chemosphere, 46: 1235~1245.

Dai C, Tian L, Zhao Y, et al. 2010. Degradation of phenanthrene by the endophytic fungus *Ceratobasidum stevensii* found in *Bischofia polycarpa*. Biodegradation, 21: 245~255.

Davison J. 1988. Plant beneficial bacteria. Nature Biotechnol, 6: 282~286.

Dean-Ross D, Cerniglia C E. 1996. Degradation of pyrene by *Mycobacterium flavescens*. Appl Microbiol Biotechnol, 46: 307~312.

Dean-Ross D, Moody J D, Freeman J P, et al. 2001. Metabolism of anthracene by a *Rhodococcus* speicies. FEMS Microbial Lett., 204: 205~211.

Denaix L, Semlali R M, Douay F. 2001. Dissolved and colloidal transport of Cd, Pb, and Zn in a silt loam soil affected by atmospheric industrial deposition. Environ. Pollut., 114: 29~38.

Derudi M, Venturini G, Lombardi G. 2007. Biodegradation combined with ozone for the remediation of contaminated soils. Eur. J. Soil. Biol., 43: 297~303.

Drever J I, Stillings L L. 1997. The role of organic acids in mineral weathering. Colloids Surf. A., 120: 167~181.

Edwards W, Bownes R, Leukes W D, et al. 1999. A capillary membrane bioreactor using immobilized polyphenol oxidase for the removal of phenols from industrial effluents. Enzyme Microb. Technol., 24: 209~217.

Elfick A P, Green S M, Pinder I M, et al. 2000. A novel technique for the detailed size characterization of wear debris. J. Mater Sci.: Mater Med, 11: 267~271.

Elgh-Dalgren K, Arwidsson Z, Camdzija A, et al. 2009. Laboratory and pilot scale soil washing of PAH and arsenic from a wood preservation site: changes in concentration and toxicity. J. Hazard Mater, 172: 1033~1040.

Eschenbach A, Weinberg R, Mahro B. 1998. Fate and stability of nonextractable residues of [^{14}C]PAH in contaminated soils under environmental stress conditions. Environ. Sci. Technol., 32: 2585~2590.

Fismes J, Perrin-Ganier C, Empereur-Bissonnet P, et al. 2002. Soil-to-root transfer and translocation of polycyclic aromatic hydrocarbons by vegetables grown on industrial contaminated soils. J. Environ. Qual., 31: 1649~1656.

Folberth C, Scherb H, Suhadolc M, et al. 2009. In situ mass distribution quotient (iMDQ) - A new factor to compare bioavailability of chemicals in soils? Chemosphere, 75: 707~713.

Forbes P J, Black K E, Hooker J E. 1997. Temperature-induced alteration to root longevity in *Lolium Perenne*. Plant Soil, 190: 87~90.

Fryer M E, Collins C D. 2003. Model intercomparition for the uptake of organic chemicals by plants. Environ. Sci. Technol.; 37: 1617~1624.

Führ F, Mittelstaedt W. 1980. Plant experiment on the bioavailability of un-extracted carbonyl-^{14}C methabenzthiazuron residues from soil. J. Agric. Food Chem., 28: 122~125.

Führ F, Opho H, Burauel P, et al. 1998. Modification of the definition of bound residues//Pesticide Bound Residues in Soil (Report of a Workshop, September 3-4, 1996). State Commission for the Assessment of Chemicals used in Agriculture. Report No. 2. Publ., Wiley-VCH.

Gao Y Z, Cao X Z, Kang F X, et al. 2011b. PAHs pass through the cell wall and partition into organelles of arbuscular mycorrhizal roots of ryegrass. J. Environ. Qual., 40: 653~656.

Gao Y Z, Cheng Z X, Ling W T, et al. 2010c. Arbuscular mycorrhizal fungal hyphae contribute to the uptake of polycyclic aromatic hydrocarbons by plant roots. Bioresour. Technol., 101: 6895~6901.

Gao Y Z, Collins C D. 2009. Uptake pathways of polycyclic aromatic hydrocarbons in white clover. Environ. Sci. Technol., 43: 6190~6195.

Gao Y Z, He J Z, Ling W T, et al. 2003. Effects of organic acids on copper and cadmium desorption from contaminated soils. Environ. Inter., 29: 613~618.

Gao Y Z, Li H, Gong S S. 2012. Ascorbic acid enhances the accumulation of polycyclic aromatic hydrocarbons in roots of tall fescue (*Festuca arundinacea* Schreb.). PloS. One, 7:e50467.

Gao Y Z, Li Q L, Ling W T, et al. 2011c. Arbuscular mycorrhizal phytoremediation of soils contaminated with phenanthrene and pyrene. J. Hazard Mater, 185: 703~709.

Gao Y Z, Ling W T. 2006. Comparison for plant uptake of phenanthrene and pyrene from soil and water. Biol. Fert. Soils, 42: 387~394.

Gao Y Z, Ling W T, Wong M H. 2006b. Plant-accelerated dissipation of phenanthrene and pyrene from water in the presence of a nonionic-surfactant. Chemosphere, 63: 1560~1567.

Gao J P, Maguhn J, Spitzauer P, et al. 1998. Sorption of pesticides in the sediment of the Teufelsweiher pond (Southern Germany). 2: Compettive adsorption, desorption of aged residues and effect of dissolved organic carbon. Water. Res., 32 : 2089~2094.

Gao Y Z, Ren L L, Ling W T, et al. 2010. Desorption of phenanthrene and pyrene in soils by root exudates. Bioresour. Technol., 101: 1159~1165.

Gao Y Z, Ren L L, Ling W T, et al. 2010b. Effects of low-molecular-weight organic acids on sorption-desorption of phenanthrene in soils. Soil Sci. Soc. Am. J., 74: 51~59.

Gao Y Z, Shen Q, Ling W T, et al. 2008b. Uptake of polycyclic aromatic hydrocarbons by *Trifolium pretense* L. from water in the presence of a nonionic surfactant. Chemosphere, 72: 636~643.

Gao Y Z, Wang N, Li H, et al. 2015.Low-molecular-weight organic acids influence the sorption of phenanthrene by different soil particle size fractions. J. Environ. Qual., 44: 219~227.

Gao Y Z, Wang Y Z, Zeng Y C, et al. 2013. Phytoavailability and rhizospheric gradient distribution of bound-PAH residues in soils. Soil Sci. Soc. Am. J., 77: 1572~1583.

Gao Y Z, Xiong W, Ling W T, et al. 2006. Sorption of phenanthrene by contaminated soils with heavy metals. Chemosphere, 65: 1355~1361.

Gao Y Z, Xiong W, Ling W T, et al. 2007. Impact of exotic and inherent dissolved organic matter on phenanthrene sorption by soils. J. Hazard. Mater., 140: 138~144.

Gao Y Z, Xiong W, Ling W T, et al. 2008. Partitioning of polycyclic aromatic hydrocarbons between plant roots and water. Plant Soil, 311: 201~209.

Gao Y Z, Yang Y, Ling W T, et al. 2011. Gradient distribution of root exudates and polycyclic aromatic hydrocarbons in rhizosphere soil. Soil Sci. Soc. Am. J., 75:1694~1703.

Gao Y Z, Yuan X J, Lin X H, et al. 2015b. Low-molecular-weight organic acids enhance the release of bound PAH residues in soil. Soil Till. Res., 145: 103~110.

Gao Y Z, Zeng Y C, Shen Q, et al. 2009. Fractionation of polycyclic aromatic hydrocarbon residues in soils. J. Hazard Mater, 172:897~903.

Gao Y Z, Zhang Y, Liu J, et al. 2013b. Metabolism and subcellular distribution of anthracene in tall fescue (*Festuca arundinacea* Schreb.). Plant Soil, 365: 171~182.

Gao Y Z, Zhu L Z. 2003. Phytoremediation and its models for organic contaminated soils. J. Environ. Sci., 15: 302~310.

Gao Y Z, Zhu L Z. 2004. Plant uptake, accumulation and translocation of phenanthrene and pyrene in soils. Chemosphere, 55: 1169~1178.

Gao Y Z, Zhu L Z. 2005. Phytoremediation of phenanthrene and pyrene in soils. J. Environ. Sci., 16: 14~18.

Gao Y Z, Zhu L Z, Ling W T. 2005. Application of the partition-limited model for plant uptake of organic chemicals from soil and water. Sci. Total Environ., 336: 171~182.

Garbeva P, Overbeek V L, Vuurde V J, et al. 2001. Analysis of endophytic bacterial communities of potato by plating and denaturing gradient gel electrophoresis (DGGE) of 16S rDNA based PCR fragments. Microbial Ecol., 41: 369~383.

George M, Julia A B, Timothy G L. 2004. Taxonomic outline of the prokaryotes Bergey's manual of systematic bacteriology, 2nd edition. New York: Springer-Verlag.

Germaine K, Keogh E, Garcia-Cabellos G, et al. 2004. Colonisation of poplar trees by *gfp* expressing bacterial endophytes. FEMS Microbiol Ecol., 48: 109~118.

Germaine K J, Keogh E, Ryan D, et al. 2009. Bacterial endophyte-mediated naphthalene phytoprotection and phytoremediation. FEMS Microbiol. Lett., 296: 226~234.

Germaine K J, Liu X M, Cabellos G G, et al. 2006. Bacterial endophyte-enhanced phytoremediation of the organochlorine herbicide 2,4- dichlorophenoxyacetic acid. FEMS Microbiol. Ecol., 57: 302~310.

Gevao B, Semple K T, Jones K C. 2000. Bound pesticide residues in soils: a review. Environ. Pollut., 108: 3~14.

Glover L J, Eick M J, Brady P V. 2002. Desorption kinetics of Cd^{2+} and Pb^{2+} from goethite: influence of time and organic acids. Soil Sci. Soc. Am. J., 66: 797~804.

Gomez-Eyles J L, Collins C D, Hodson M E. 2010. Relative proportions of polycyclic aromatic hydrocarbons differ between accumulation bioassays and chemical methods to predict bioavailability. Environ. Pollut., 158: 278~284.

Gomez-Eyles J L, Collins C D, Hodson M E. 2011. Using deuterated PAH amendments to validate chemical extraction methods to predict PAH bioavailability in soils. Environ. Pollut., 159: 918~923.

Gong S S, Han J, Gao Y Z, et al. 2011. Effects of inhibitor and safener on enzyme activity and phenanthrene metabolism in root of tall fescue. Acta. Ecologica. Sinica., 31: 4027~4033 (in Chinese).

Grayston S J, Vaughan D, Jones D. 1997. Rhizosphere carbon flow in trees, in comparison with annual plants: the importance of root exudation and its impact on microbial activity and nutrient availability. Appl. Soil Ecol., 5: 29~56.

Greenwood S J, Rutter A, Zeeb B A. 2011. The absorption and translocation of polychlorinated biphenyl congeners by *Cucurbita pepo ssp pepo*. Environ. Sci. Technol., 45: 6511~6516.

Gronwal J W, Connelly J A. 1991. Effect of monooxygenase inhibitors on bentazon uptake and metabolism in maize cell suspension cultures. Pestic. Biochem. Physiol., 40: 284~294.

Guthrie E A, Pfaender F K. 1998. Reduced pyrene bioavailability in microbially active soils. Environ.

Sci. Technol., 32: 501~508.

Gunasekara A S, Simpson M J, Xing B. 2003. Identification and characterization of sorption domains in soil organic matter using structurally modified humic acid. Environ Sci Technol, 37: 852~858.

Haby P A, Crowley D E. 1996. Biodegradation of 3-chlorobenzoate as affected by rhizodeposition and selected carbon substrates. J. Environ. Qual., 25 : 304~310.

Haftka J J, Govers H A, Parsons J R. 2010. Influence of temperature and origin of dissolved organic matter on the partitioning behavior of polycyclic aromatic hydrocarbons. Environ. Sci. Pollut. Res., 17: 1070~1079.

Hannink N K, Rosser S J, Bruce N C. 2002. Phytoremediation of explosives. Crit. Rev. Plant Sci., 21: 511~538.

Hao Z B, Cang J, Xu Z. 2004. Experiments of plant physiology. Haer-bin industrial university publishing company: 115~116.

Harish S, Kavino M, Kumar N, et al. 2008. Biohardening with plant growth promoting rhizosphere and endophytic bacteria induces systemic resistance against *banana bunchy top virus*. Appl. Soil. Ecol., 39: 187~200.

Harish S, Kavino M, Kumar N, et al. 2009. Induction of defense-related proteins by mixtures of plant growth promoting endophytic bacteria against Banana bunchy top virus. Biol. Control, 51: 16~25.

Harms H H. 1996. Bioaccumulation and metabolic fate of sewage derived organic xenobiotics in plants. Sci. Total Environ., 185: 83~92.

Harrad S, Ren J Z, Hazrati S. 2006. Chiral signatures of PCBs 95 and 149 in indoor air, grass, duplicate diets and human faeces. Chemosphere, 63: 1368~1376.

Hart M M, Trevors J T. 2005. Microbe management: application of mycorrhyzal fungi in sustainable agriculture. Front Ecol. Environ., 3: 533~539.

Hatzinger P B, Alexander M. 1995. Effect of aging of chemicals in soil on their biodegradability and extractability. Environ. Sci. Technol., 29: 537~545.

He Y, Xu J M, Lv X F, et al. 2009. Does the depletion of pentachlorophenol in root–soil interface follow a simple linear dependence on the distance to root surfaces? Soil Biol. Biochem., 41: 1807~1813.

Hees P A W, Andersson A M T, Lundstrom U S. 1996. Separation of organic low molecular weight aluminum complexes in soil solution by liquid chromatography. Chemosphere, 33: 1951~1966.

Henriksen A, Smith A T, Gajhede M. 1999. The structures of the horseradish peroxidase c-ferulic acid complex and the ternary complex with cyanide suggest how peroxidases oxidize small phenolic substrates. J. Biol. Chem., 274: 35005~35011.

Hildebrandt U, Regvar M, Bothe H. 2007. Arbuscular mycorrhiza and heavy metal tolerance.

Phytochem., 68: 139~146.

Ho Y N, Shih C H, Hsiao S C, et al. 2009. A novel endophytic bacterium, *Achromobacter xylosoxidans*, helps plants against pollutant stress and improves phytoremediation. J. Biosci. Bioeng., 108: S75~S95.

Hodge A, Stewart J, Robinson D, et al. 2000. Competition between roots and soil micro-organisms for nutrients from nitrogen-rich patches of varying complexity. J. Ecol., 88: 150~164.

Hofrichter M, Scheibner K, Schneegass I, et al. 1998. Enzymatic combustion of aromatic and aliphatic compounds by manganese peroxidase from *Nematoloma frowardii*. Appl. Environ. Microbiol., 64: 399~404.

Horinouchi M, Nishio Y, Shimpo E. 2000. Removal of polycyclic aromatic hydrocarbons from oil-contaminated Kuwaiti soil. Biotechnol. Lett., 22: 687~691.

Howsam M, Jones K C, Ineson P. 2001. PAHs associated with the leaves of three deciduous tree species. II: uptake during a growing season. Chemosphere, 44: 155~164.

Hsu F C, Marxmiller R L, Yang A Y S. 1990. Study of root uptake and xylem translocation of cinmethylin and related-compounds in detopped soybean roots using a pressure chamber technique. Plant Physiol., 93: 1573~1578.

Huang W H, Keller W D. 1972. Organic acids as agents of chemicals weathering of silicate minerals. Nature, 239: 149~151.

Huang H L, Zhang S Z, Shan X Q, et al. 2007. Effect of arbuscular mycorrhizal fungus (*Glomus caledonium*) on the accumulation and metabolism of atrazine in maize (*Zea mays* L.) and atrazine dissipation in soil. Environ. Pollut., 146: 452~457.

Hückelhoven R, Schuphan I, Thiede B, et al. 1997. Biotransformation of pyrene by cell cultures of soybean (*Glycine max* L.), wheat (*Triticum aestivum* L.), Jimsonweed (*Datura stramonium* L.), and purlple foxglove (*Digitalis purpurea* L.). J. Agric. Food Chem., 45: 263~269.

Hwang S, Cutright T J. 2004. Preliminary evaluation of PAH sorptive changes in soil by Soxhlet extraction. Environ. Int., 30: 151~158.

Institute of Soil Science, Chinese Academy of Sciences (ISSCAS). 1980. Chinese Soil. Science Press Ltd., Beijing.

Jackson D, Garrett B, Bishop B. 1984. Comparison of batch and column methods for assessing leachability of hazardous waste. Environ. Sci. Technol., 18: 668~673.

Jagadamma S, Mayes M A, Zinn Y L, et al. 2014. Sorption of organic carbon compounds to the fine fraction of surface and subsurface soils. Geoderma, 213: 79~86.

James E K, Gyaneshwar P, Mathan N, et al. 2002. Infection and colonization of rice seedlings by the plant growth-promoting bacterium *Herbaspirillum seropedicae* Z67. Molec. Plant-Microbe Interact., 15: 894~906.

Joner E J, Corgiés, Amellal N, et al. 2002. Nutritional constraints to degradation of polycyclic aromatic hydrocarbons in a simulated rhizosphere. Soil. Biol. Biochem., 34: 859~864.

Joner E J, Johansen A, dela Cruz M A T, et al. 2001. Rhizosphere effects on microbial community structure and dissipation and toxicity of polycyclic aromatic hydrocarbons (PAHs) in spiked soil. Environ. Sci. Technol., 35: 2773~2777.

Joner E J, Leyval C. 2003. Rhizosphere gradients of polycyclic aromatic hydrocarbon (PAH) dissipation in two industrial soils and the impact of arbuscular mycorrhiza. Environ. Sci. Technol., 37: 2371~2375.

Jones D L, Brassington D S. 1998. Sorption of organic acids in acid soils and its implications in the rhizosphere. Europ. J. Soil Sci., 49: 447~455.

Jones D L, Hodge A, Kuzyakov Y. 2004. Plant and mycorrhizal regulation of rhizodeposition. New Phytol., 163: 459~480.

Jones K D, Tiller C L. 1999. Effect of solution chemistry on the extent of binding of phenanthrene by a soil humic acid: a comparison of dissolved and clay bound humic acids. Environ. Sci. Technol., 33: 580~587.

Jerzy D, Konrad H, Rangaswamy V. 1997. Formation of soil-bound residues of cyprodinil and their plant uptake, J. Agric. Food Chem., 45: 514~520.

Juhasz A L, Naidu R. 2000. Bioremediation of high molecular weight polycyclic aromatic hydrocarbons: a review of the microbial degradation of benzo[a]pyrene. Int. Biodeter. Biodegrad., 42: 57~88.

Käcker T, Haupt E T K, Garms C, et al. 2002. Structural characterisation of humic acid-bound PAH residues in soil by ^{13}C-CPMAS-NMR-spectroscopy: evidence of covalent bonds. Chemosphere, 48: 117~131.

Kang F X, Chen D S, Gao Y Z, et al. 2010. Distribution of polycyclic aromatic hydrocarbons in subcellular root tissues of ryegrass (*Lolium multiflorum* Lam.). BMC Plant Biol., 10:210.

Kaplan C W, Kitts C L. 2004. Bacterial succession in a petroleum land treatment unit. Appl Environ. Microbiol., 70: 1777~1786.

Kapley A, Siddiqui S, Misra K, et al. 2007. Preliminary analysis of bacterial diversity associated with the Porites coral from the Arabian sea. World J. Microb. Biot., 23: 923~930.

Karickhoff S W, Brown D S, Scott T A. 1979. Sorption of hydrophobic pollutants on natural sediments. Wat. Res., 13: 241~248.

Karim Z, Husain Q. 2010. Removal of anthracene from model wastewater by immobilized peroxidase from *Momordica charantia* in batch process as well as in a continuous spiral-bed reactor. J. Mol. Catal. B-Enzym., 66: 302~310.

Kastner M, Breuer-Jammali M, Mahro B. 1998. Impact of inoculation protocols, salinity and pH on the degradation of polycyclic aromatic hydrocarbons (PAHs) and survival of PAH-degrading bacteria introduced into soil. Appl. Environ. Microbiol., 64: 359~362.

Kelsey J W, Kottler B D, Alexander M. 1997. Selective chemical extractants to predict bioavailability of soil-aged organic chemicals. Environ. Sci. Technol., 31: 214~217.

Khan Z, Doty S. 2011. Endophyte-assisted phytoremediation. Curr. Top Plant Biol., 12: 97~105.

Khan S U, Ivarson K C. 1981. Microbiological release of unextracted residues from an organic soil treated with prometryn. J. Agric. Food Chem., 29: 1301~1311.

Khan A G, Kuek C, Chaudhry T M, et al. 2000. Role of plants, mycorrhizae and phytochelators in heavy metal contaminated land remediation. Chemosphere, 41: 197~207.

Kipopoulou A M, Manoli E, Samara C. 1999. Bioconcentration of PAHs in vegetables grown in an industrial area. Environ. Pollut., 106: 369~380.

Kohl S D, Rice J A. 1998. The binding of contaminants to humin: a mass balance. Chemosphere, 36: 251~261.

Köller G, Möder M, Czihal K. 2000. Peroxidative degradation of selected PCB: a mechanistic study. Chemosphere, 41: 1827~1834.

Kong H L, Sun, R Gao Y Z, et al. 2013. Elution of polycyclic aromatic hydrocarbons in soil columns using low-molecular-weight organic acids. Soil. Sci. Soc. Am. J., 77: 72~82.

Kozdroj J, Van Elsas J D. 2000. Response of the bacterial community to root exudates in soil polluted with heavy metals assessed by molecular and cultural approaches. Soil Biol. Biochem., 32: 1405~1417.

Kpomblekou-A K, Tabatabai M A. 2003. Effect of low-molecular weight organic acids on phosphorus release and phytoavailability of phosphorus in phosphate rocks added to soils. Agric. Ecosys. Environ., 100: 275~284.

Krauss M, Wilcke W. 2005. Persistent organic pollutants in soil density fractions: distribution and sorption strength. Chemosphere, 59: 1507~1515.

Krishnamurti G S R, Cieslinski G, Huang P M, et al. 1997. Kinetics of cadmium release from soils as influenced by organic acids: implication in cadmium availability. J. Environ. Qual., 26: 271~277.

Lai Y, Wang Q, Yang L, et al. 2006. Subcellular distribution of rare earth elements and characterization of their binding species in a newly discovered hyperaccumulator *Pronephrium simplex*. Talanta, 70: 26~31.

Launen L A, Pinto L J, Percival P W, et al. 2000. Pyrene is metabolized to bound residues by *Penicillium janthinellum* SFU403. Biodegrad, 11: 305~312.

Leahy J G, Colwell R R. 1990. Microbial degradation of hydrocarbons in the environment. Microbiol. Rev., 54: 305~315.

Lesan H M, Bhandari A. 2004. Contact-time-dependent atrazine residue formation in surface soils. Water Res., 38: 4435~4445.

LeFevre G H, Hozalski R M, Novak P J. 2013. Root exudate enhanced contaminant desorption: an abiotic contribution to the rhizosphere effect. Environ. Sci. Technol., 47: 11545~11553.

Li H, 2006. Plant physiology (The second edition). Beijing: High Education Press.

Li H, Sheng G, Chiou C T, et al. 2005. Relation of organic contaminant equilibrium sorption and kinetic uptake in plants. Environ. Sci. Technol., 39: 4864~4870.

Li H, Sheng G, Sheng W, et al. 2002. Uptake of trifluralin and lindane from water by ryegrass. Chemosphere, 48: 335~341.

Li H, Teppen B J, Johnston C T, et al. 2004. Thermodynamics of nitroaromatic compound adsorption from water by smectite clay. Environ. Sci. Technol., 38: 5433~5442.

Li J, Xu R, Tiwari D, et al. 2006. Effect of low-molecular-weight organic acids on the distribution of mobilized Al between soil solution and solid phase. Appl. Geochem., 21: 1750~1759.

Li T Q, Yang X E, Yang J Y, et al. 2006b. Zn accumulation and subcellular distribution in the Zn hyperaccumulator Sedurn alfredii hance. Pedosphere, 16: 616~623.

Li Y, Yediler A, Ou Z, et al. 2001. Effects of a non-ionic surfactant (Tween80) on the mineralization, metabolism and uptake of phenanthrene in wheat-solution-lava microcosm. Chemosphere, 45: 67~75.

Lin Q, Chen Y X, Chen H M, et al. 2000. The ecological effects of Pb and Cd on the root activities of wheat. Acta. Ecologica. Sinica., 20: 634~638 (In Chinese).

Ling W T, Dang H J, Liu J. 2013. In situ gradient distribution of polycyclic aromatic hydrocarbons (PAHs) in contaminated rhizosphere soil: a field study. J. Soils. Sedim., 13: 677~685.

Ling W T, Gao Y Z. 2004. Promoted dissipation of phenanthrene and pyrene in soils by amaranth (*Amaranthus tricolor* L.). Environ. Geol., 46: 553~560.

Ling W T, Lu X D, Gao Y Z, et al. 2012. Polyphenol oxidase activity in subcellular fractions of tall fescue (*Festuca arundinacea* Schreb.) contaminated by polycyclic aromatic hydrocarbons. J. Environ. Qual., 41: 807~813.

Ling W T, Zeng Y C, Gao Y Z, et al. 2010. Availability of polycyclic aromatic hydrocarbons in aging soils. J. Soils Sedim., 10: 799~807.

Ling W T, Ren L L, Gao Y Z, et al. 2009. Impact of low-molecular-weight organic acids on the availability of phenanthrene and pyrene in soil. Soil Biol. Biochem., 41: 2187~2195.

Ling W T, Sun R, Gao X, et al. 2015. Low-molecular-weight organic acids enhance desorption of polycyclic aromatic hydrocarbons from soil. Europ. J. Soil Sci., 66: 339~347.

Ling W T, Xu J M, Gao Y Z. 2006. Dissolved organic matter-enhanced sorption of atrazine by soil in three-phase system. Biol. Fert. Soil., 42: 418~425.

Ling W T, Xu J M, Wang H Z, et al. 2005. Sorption of dissolved organic matter and its effects on the atrazine sorption on soils. J. Environ. Sci., 16: 478~482.

Liste H H, Alexander M. 2000. Plant-promoted pyrene degradation in soils. Chemosphere, 40: 7~10.

Liste H H, Alexander M. 2002. Butanol extraction to predict bioavailability of PAHs in soil. Chemophere, 46: 1011~1017.

Liu J, Liu S, Sun K, et al. 2014. Colonization on root surface by a phenanthrene-degrading endophytic bacterium and its application for reducing plant phenanthrene contamination. PLoS ONE, 9: e108249.

Lodewyckx C, Vangronsveld J, Porteous F, et al. 2002. Endophytic bacteria and their potential applications. Crit. Rev. in Plant Sci., 21: 583~606.

Loiseau L, Barriuso E. 2002. Characterization of the atrazine's bound (non-extractable) residues using fractionation techniques for soil organic matter. Environ. Sci. Technol., 36: 683~689.

Louch,X, Voltz M. 2007. Aging effects on the availability of herbicides to runoff transfer. Environ. Sci. Technol., 41: 1137~1144.

Lozano-Rodriguez E, Hernandez L, Bonay P, et al. 1997. Distribution of cadmium in shoot and root tissues of maize and pea plants: physiological disturbances. J. Exper. Bot., 48 : 123~128.

Lu X D, Gao Y Z, Ling W T, et al. 2008. Effects of polycyclic aromatic hydrocarbons on POD and PPO in *Lolium multiflorum* Lam. J. Agro-Environ. Sci., 27: 1969~1973 (In Chinese).

Lu H L, Yan C L, Liu J C. 2007. Low-molecular-weight organic acids exuded by Mangrove (*Kandelia candel* (L.) Druce) roots and their effect on cadmium species change in the rhizosphere. Environ. Exper. Bot., 61: 159~166.

Lynch J M, Whipps J M. 1990. Substrate flow in the rhizosphere. Plant Soil, 129: 1~10.

Macek T, Kotrba P, Svatos A, et al. 2008. Novel roles for genetically modified plants in environmental protection. Trends Biotechnol., 26: 146~152.

Mackova M, Macek T, Ocenaskova J, et al. 1997. Biodegradation of polychlorinated biphenyls by plant cells. Inter. Biodeter. Biodegrad., 39: 317~325.

Macleod C J A, Semple K T. 2003. Sequential extraction of low concentrations of pyrene and formation of non-extractable residues in sterile and non-sterile soils. Soil Biol. Biochem., 35: 1443~1450.

Mattina M J I, Lannucci-Berger W, Musante C, et al. 2003. Concurrent plant uptake of heavy metals and persistent organic pollutants from soils. Environ. Pollut., 124: 375~378.

Maxin C R, Kögel-Knabner I. 1995. Partitioning of polycyclic aromatic hydrocarbons (PAHs) to water-soluble soil organic matter. Europ. J. Soil. Sci., 46: 193~204.

Mayer A M. 2006. Polyphenol oxidases in plants and fungi: Going places? A review. Phytochem., 67: 2318~2331.

McCully M E. 2001. Niches for bacterial endophytes in crop plants: a plant biologist's view Aust. Aust. J. Plant. Physiol., 28: 983~990.

McGinley P M, Katz L E, et al. 1993. A distributed reactivity model for sorption by soils and sediments. 2. Multicomponent systems and competitive effects. Environ. Sci. Technol., 27: 1524~1532.

Meharg A A, Killham K. 1990. Carbon distribution within the plant and rhizosphere for *Lolium perenne* subjected to anaerobic soil conditions. Soil Biol. Biochem., 22: 643~647.

Mendes R, Kleiner A A P, Araujo W L, et al. 2007. Diversity of cultivated endophytic bacteria from sugarcane: genetic and biochemical characterization of *Burkholderia cepacia* complex isolates. Appl. Environ. Microbiol., 73: 7259~7267.

Moen M A, Hammel K E. 1994. Lipid peroxidation by the manganese peroxidase of *Phanerochaete chrysosporium* is the basis for phenanthrene oxidation by the intact fungus. Appl. Environ. Microbiol., 60: 1956~1961.

Monteiro R T R, Queiroz B P V, Cella A C. 1999. Distribution of ^{14}C-atrazine residues in humus fractions of a Brazilian soil. Chemosphere, 39: 293~301.

Moore F P, Barac T, Borremans B, et al. 2006. Endophytic bacterial diversity in poplar trees growing on a BTEX-contaminated site: The characterisation of isolates with potential to enhance phytoremediation. Syst. Appl. Microbiol., 29: 539~556.

Mott H V. 2002. Association of hydrophobic organic contaminants with soluble organic matter: evaluation of the database of K_{doc} values. Adv. Environ. Res., 6: 577~593.

Muratova A, Golubev S, Wittenmayer L, et al. 2009. Effect of the polycyclic aromatic hydrocarbon phenanthrene on root exudation of *Sorghum bicolor* (L.) Moench. Environ. Exper. Bot., 66: 514~521.

Muratova A Y, Kapitonova V V, Chernyshova M P, et al. 2009b. Enzymatic activity of alfalfa in a phenanthrene-contaminated environment. World Acad. Sci. Eng. Technol., 58: 569~574.

Murphy E M, Zachara J M. 1995. The role of sorbed humic substances on the distribution of organic and inorganic contaminants in groundwater. Geoderma, 67: 103~124.

Murphy E M, Zachara J M, Smith S C. 1990. Influence of mineral-bound humic substances on the sorption of hydrophobic organic compounds. Environ. Sci. Technol., 24: 1507~1516.

Nam K, Chung N, Alexander M. 1998. Relationship between organic matter content of soil and the sequestration of phenanthrene. Environ. Sci. Technol., 32: 3785~3788.

Nardi S, Reniero F, ConcheriG. 1997. Soil organic matter mobilization by root exudates of three maize hybrids. Chemosphere, 35: 2237~2244.

Nardi S, Tosoni M, Pizzeghello D, et al. 2005. Chemical characteristics and biological activity of organic substances extracted from soils by root exudates. Soil Sci. Soc. Am. J., 69: 2012~2019.

Nelson S D, Letey J, Farmer W J, et al. 1998. Facilitated transport of napropamide by dissolved organic matter in sewage sludge-amended soil. J. Environ. Qual., 27: 1194~1200.

Nemeth-Konda L, Fűleky Gy, Morovjan Gy, et al. 2002. Sorption behaviour of acetochlor, atrazine, carbendazim, diazinon, imidacloprid and isoproturon on Hungarian agricultural soil. Chemosphere, 48: 545~552.

Nepovim A, Hubalek M, Podlipna R, et al. 2004. In-vitro degradation of 2,4,6-trinitrotoluene using plant tissue cultures of *Solanum aviculare* and *Rheum palmatum*. Eng. Life Sci., 4: 46~49.

Newman L A, Reynolds C M. 2005. Bacteria and phytoremediation: New uses for endophytic bacteria in plants. Trends. Biotechnol., 23: 6~8.

N'Guessan A L, Alderman N S, O'Connor K, et al. 2006. Peroxy-acid treatment of selected PAHs in sediments. International J. Environ. Waste Manag., 1: 61~74.

Nieman J K, Sims R C, Sims J L, et al. 1999. [^{14}C]Pyrene bound residue evaluation using MIBK fractionation method for creosote-contaminated soil. Environ. Sci. Technol., 33: 776~781.

Nkedi-Kizza P, Rao P S C, Hornsby A G. 1987. Influence of organic cosolvents on leaching of hydrophobic organic chemicals through soils. Environ. Sci. Technol., 21: 1107~1111.

Nogales B, Moore E R B, Llobet-Brossa E, et al. 2001. Combined use of 16S ribosomal DNA and 16S rRNA to study the bacterial community of polychlorinated biphenyl-polluted soil. Appl. Environ. Microbiol., 67: 1874~1884.

Noordkamp E R, Grotenhuis J T C, Rulkens W H. 1997. Selection of an efficient extraction method for the determination of polycyclic aromatic hydrocarbons in contaminated soil and sediment. Chemosphere, 35:1907~1917.

Northcott G L, Jones K C. 2000. Experimental approaches and analytical techniques for determining organic compound bound residues in soil and sediment. Environ. Pollut., 108: 19~43.

Nowak K M, Miltner A, Gehre M, et al. 2011. Formation and fate of bound residues from microbial biomass during 2,4-D degradation in soil. Environ. Sci. Technol., 45: 999~1006.

Nunan N, Daniell T J, Singh B K, et al. 2005. Links between plant and rhizoplane bacterial communities in grassland soils characterized using molecular techniques. Appl. Environ. Microbiol., 71: 6784~6792.

Ouvrard S, Lapole D, Morel J L. 2006. Root exudates impact on phenanthrene availability. Water Air Soil Pollut., 6: 343~352.

Overbeek L S, Elsas J D. 1995. Root exudates-induced promoter activity in pseudomonas fluorenscens mutants in the wheat rhizosphere. Appl. Environ. Microbiol., 3: 890~898.

Paterson S, Mackay D. 1994. A model of organic chemical uptake by plants from soil and the atmosphere. Environ. Sci. Technol., 28: 2259~2266.

Paterson E, Sim A. 2000. Effect of nitrogen supply and defoliation on loss of organic compounds from roots of *Festuca rubra*. J. Exp. Bot., 51: 1449~1457.

Peng A P, Liu J, Ling W T, et al. 2015. Diversity and distribution of 16S rRNA and phenol monooxygenase genes in the rhizosphere and endophytic bacteria isolated from PAH-contaminated sites. Sci. Rep., 5:12173.

Peng, Liu J, Gao Y Z, et al. 2013. Distribution of endophytic bacteria in *Alopecurus aequalis* Sobol and *Oxalis corniculata* L. from soils contaminated by polycyclic aromatic hydrocarbons. PLoS ONE, 8:e83054.

Pernot A, Ouvrard S, Leglize P, et al. 2013. Protective role of fine silts for PAH in a former industrial soil. Environ. Pollut., 179: 81~87.

Petersen L S, Larsen E H, Larsen P B, et al. 2002. Uptake of trace elements and PAHs by fruit and vegetables from contaminated soils. Environ. Sci. Technol., 36: 3057~3063.

Petra M, David C, Ching H Y. 2004. Development of specific rhizosphere bacterial communities in relation to plant species, nutrition and soil type. Plant Soil, 261: 199~208.

Phillips L A, Germida J J, Farrell R E. 2008. Hydrocarbon degradation potential and activity of endophytic bacteria associated with prairie plants. Soil Biol. Biochem., 40: 3054~3064.

Phillips D A, Ferris H, Cook D R, et al. 2003. Molecular control points in rhizosphere food webs. Ecol., 84: 816~826.

Pier M D, Zeeb B A, Reimer K J. 2002. Patterns of contamination among vascular plants exposed to local sources of polychlorinated biphenyls in the Canadian Arctic and Subarctic. Sci. Total Environ., 297: 215~227.

Pignatello J J, Xing B. 1996. Mechanisms of slow sorption of organic chemicals to natural particles. Environ. Sci. Technol., 30: 1~11.

Pignatello J J. 1998. Soil organic matter as a nanoporous sorbent of organic pollutants. Adv. Colloid. Interf. Sci., 76-77: 445~467.

Polder M D, Hulzebos E M, Jager D T. 1995. Validation of models on uptake of organic chemicals by plant roots. Environ. Toxicol. Chem.; 14: 1615~1623.

Pritchina O, Ely C, Smets B F. 2011. Effects of PAH-contaminated soil on rhizosphere microbial communities. Water Air Soil Pollut., 222: 17~25.

Prothero D R, Schwab F. 1996, Sedimentary geology: An introduction to sedimentary rocks and stratigraphy: W. H. Freeman and Company, New York.

Qin F, Shan X, Wei B. 2004. Effects of low-molecular-weight organic acids and residence time on desorption of Cu, Cd, and Pb from soils. Chemosphere, 57: 253~263.

Qing W, Zhao X, Zhao S Y. 2006. Application of PCR-DGGE in research of bacterial diversity in drinking water. Biomed. Environ. Sci., 19:371~374.

Raber B, Kögel-Knabner I. 1997. Influence of origin and properties of dissolved organic matter on the partition of polycyclic aromatic hydrocarbons. Europ. J. Soil Sci., 48: 443~455.

Raber B, Kögel-Knabner I, Stein C, et al. 1998. Partitioning of polycyclic aromatic hydrocarbons to dissolved organic matter from different soils. Chemosphere, 36: 79~97.

Rajkumar M, Ae N, Freitas H. 2009. Endophytic bacteria and their potential to enhance heavy metal phytoextraction. Chemosphere, 77: 153~160.

Raynaud X. 2010. Soil properties are key determinants for the development of exudate gradients in a rhizosphere simulation model. Soil Biol. Biochem., 42: 210~219.

Rehmann K, Noll H P, Steiberg C E W, et al. 1998. Pyrene degradation by *Mycobacterium* sp. Strain KR2. Chemosphere, 36: 2977~2992.

Reid B J, Stokes J D, Jones K C, et al. 2000. Nonexhaustive cyclodextrin-based extraction technique for the evaluation of PAH bioavailability. Environ. Sci. Technol., 34: 3174~3179.

Reilley K A, Banks M K, Schwab A P. 1996. Dissipation of polycyclic aromatic hydrocarbons in the rhizosphere. J. Environ. Qual., 25: 212~219.

Rentz J A, Alvarez P J J, Schnoor J L. 2005. Benzo[a]pyrene co-metabolism in the presence of plant root extracts and exudates: Implications for phytoremediation. Environ. Pollut., 136: 477~484

Richnow H H, Annweiler E, Koning M, et al. 2000. Tracing the transformation of labelled [1-^{13}C]phenanthrene in a soil bioreactor. Environ. Pollut., 108: 91~101.

Richnow H H, Eschenbach A, Mahro B, et al. 1999. Formation of nonextractable soil residues: A stable isotope approach. Environ. Sci. Technol., 33: 3761~3767.

Richnow H H, Seifert R, Kästner M, et al. 1995. Rapid screening of PAH-residues in bioremediated soils. Chemosphere, 31:3991~3999.

Roberts T R. 1984. Non-extractable pesticide residues in soils and plants. Pure App. Chem., 56: 945~995.

Robson A S P, Leandro B, Benedict C O, et al. 2010. 19th World Congress of Soil Science, Soil Solutions for a Changing World 1~6 August 2010, Brisbane, Australia. Published on DVD.

Romano M L, Stephenson G R, Tal A, et al. 1993. The Effect of monooxygenase and glutathione s-transferase inhibitors on the metabolism of diclofop-methyl and fenoxaprop-ethyl in barley and wheat. Pestic. Biochem. Physiol., 46: 181~189.

Ryan J A, Bell R M, Davidson J M, et al. 1988. Plant uptake of non-ionic organic chemicals from soils. Chemosphere, 17: 2299~2323.

Ryan R P, Germaine K, Franks A, et al. 2008. Bacterial endophytes: Recent developments and applications. FEMS Microbiol. Lett., 278: 1~9.

Sabate J, Vinas M, Solanas A M. 2006. Bioavailability assessment and environmental fate of polycyclic aromatic hydrocarbons in biostimulated creosote-contaminated soil. Chemosphere, 63: 1648~1659.

Saby John K, Bhat S G, Prasada Rao U J S. 2011. Isolation and partial characterization of phenol oxidases from *Mangifera indica* L. sap (latex). J. Mol. Catal. B-Enzym., 68: 30~36.

Saison C, Perrin-Ganier C, Amellal S, et al. 2004. Effect of metals on the adsorption and extractability of ^{14}C-phenanthrene in soils. Chemosphere, 55: 477~485.

Sarkar S, Martinez A T, Martinez M J. 1997. Biochemical and molecular characterization of a manganese peroxidase isoenzyme from *Pleurotus ostreatus*. Biochimica Biophysica Acta, 1339: 23~30.

Schmidt M W I, Rumpel C, Kögel-Knabner I. 1999a. Particle size fractionation of soil containing coal and combusted particles. Europ. J. Soil Sci., 50: 515~522.

Schmidt M W I, Rumpel C, Kögel-Knabner I. 1999b. Evaluation of an ultrasonic dispersion procedure to isolate primary organomineral complexes from soils. Europ. J. Soil Sci., 50: 87~94.

Schroll R, Bierling B, Cao G, et al. 1994. Uptake pathways of organic chemicals from soil by agricultural plants. Chemosphere, 28: 297~303.

Semple K T, Morriss A W J, Paton G I. 2003. Bioavailability of hydrophobic organic contaminants in soils: fundamental concepts and techniques for analysis. Eur. J. Soil Sci., 54: 809~818.

Seol Y K, Lee L S. 2000. Effect of dissolved organic matter in treated effluents on sorption of atrazine and prometryn by soils. Soil Sci. Soc. Am. J., 64: 1976~1983.

Sessitsch A, Reiter B, Pfeifer U, et al. 2002. Cultivation-independent population analysis of bacterial endophytes in three potato varieties based on eubacterial and Actinomycetes-specific PCR of 16S rRNA genes. FEMS Microbiol. Ecol., 39: 23~32.

Shang T Q, Doty S L, Wilson A M, et al. 2001. Trichloroethylene oxidative metabolism in plants: the trichloroethanol pathway. Phytochem., 58:1055~1065.

Sheng X F, Chen X B, He L Y. 2008. Characteristics of an endophytic pyrene-degrading bacterium of *Enterobacter* sp. 12J1 from *Allium macrostemon Bunge*. Int. Biodeterior. Biodegrad., 62: 88~95.

Sherman T O, Vaughn K C, Duke S O. 1991. A limited survey of the phylogenetic distribution of polyphenol oxidase. Phytochem., 30: 2499~2506.

Shieh W J, Hedges A R. 1996. Properties and applications of cyclodextrins. J. Macromol. Sci. Pure Appl. Chem. A., 33: 673~683.

Shin D S, Park M S, Jung S, et al. 2007. Plant growth-promoting potential of endophytic bacteria isolated from roots of coastal sand dune plants. J. Microbiol. Biotechnol., 17: 1361~1368.

Siciliano S D, Fortin N, Mihoc A, et al. 2001. Selection of specific endophytic bacterial genotypes by plants in response to soil contamination. Appl. Environ. Microbiol., 67: 2469~2475.

Simonich S L, Hites R A. 1994. Vegetation-atmosphere partitioning of polycyclic aromatic hydrocarbons. Environ. Sci. Technol., 28: 939~943.

Simonich S L, Hites R A. 1995. Organic pollutant accumulation in vegetation. Environ. Sci. Technol.,

29: 2095~2103.

Singh B K, Millard P, Whiteley A S, et al. 2004. Unravelling rhizosphere–microbial interactions: opportunities and limitations. Trends Microbiol., 12: 386~393.

Smalla K, Wieland G, Buchner A, et al. 2001. Bulk and rhizosphere soil bacterial communities studied by denaturing gradient gel electrophoresis: plant-dependent enrichment and seasonal shifts revealed. Appl. Environ. Microbiol., 67: 4742~4751.

Smith K E C, Green M, Thomas G O, et al. 2001. Behavior of sewage sludge-derived PAHs on pasture. Environ. Sci. Technol., 35: 2141~2150.

Smith M J, Lethbridge G, Burns R G. 1999. Fate of phenanthrene, pyrene and benzo[a]pyrene during biodegradation of crude oil added to two soils. FEMS Microbiol. Lett., 173: 445~452.

Smirnoff N. 2000. Ascorbic acid: metabolism and functions of a multi-facetted molecule. Curr. Opin. Plant Biol., 3: 229~235.

Sobral J K, Araújo W L, Mendes R, et al. 2005. Isolation and characterization of endophytic bacteria from soybean (*Glycine max*) grown in soil treated with glyphosate herbicide. Plant Soil, 273: 91~99.

Somnath M, Subhankar C, Tapan K D. 2007. A novel degradation pathway in the assimilation of phenanthrene by *Staphylococcus* sp. strain PN/Y via meta-cleavage of 2-hydroxy-1-naphthoic acid: formation of trans-2, 3-dioxo-5- (29-hydroxyphenyl)- pent-4-enoic acid. Microbiology, 153: 2104~2115.

Spark K M, Swift R S. 2002. Effect of soil composition and dissolved organic matter on pesticide sorption. Sci. Total Environ., 298: 147~151.

Sterling T M, Balke N E. 1990. Bentazon uptake and metabolism by cultured plant cells in the presence of monooxygenase inhibitors and cinnamic acid. Pestic. Biochem. Physiol., 38: 66~75.

Strahm B D, Harrison R B. 2008. Controls on the sorption, desorption and mineralization of low-molecular-weight organic acids in variable-charge soils. Soil Sci. Soc. Am. J., 72: 1653~1664.

Subramaniam K, Stepp C, Pignatello J J, et al. 2004. Enhancement of polynuclear aromatic hydrocarbon desorption by complexing agents in weathered soil. Environ. Eng. Sci., 21: 515~523.

Suflita J M, Bollag J M. 1981. Polymerization of phenolic compounds by a soil–enzyme complex. Soil Sci. Soc. Am. J., 45: 297~302.

Sun H W, Li J G. 2005. Availability of pyrene in unaged and aged soils to earthworm uptake, butanol extraction and SFE. Water Air Soil. Pollut., 166: 353~365.

Sun B Q, Ling W T, Wang Y Z. 2013. Can root exudate components influence the availability of pyrene in soil. J. Soils Sedim., 13: 1161~1169.

Sun B Q, Liu J, Gao Y Z, et al. 2012. The impact of different root exudate components on phenanthrene availability in soil. Soil Sci. Soc. Am. J., 76: 2041~2050.

Sun K, Liu J, Gao Y Z, et al. 2014a. Isolation, plant colonization potential, and phenanthrene degradation performance of the endophytic bacterium *Pseudomonas* sp. Ph6-*gfp*. Sci. Rep., 4: 5462.

Sun K, Liu J, Gao Y Z, et al. 2015. Inoculating plants with the endophytic bacterium Pseudomonas sp. Ph6-gfp to reduce phenanthrene contamination. Environ Sci Pollut Res 2015. DOI 10.1007/s11356-015-5128-9

Sun K, Liu J, Jin L, et al. 2014b. Utilizing pyrene-degrading endophytic bacteria to reduce the risk of plant pyrene contamination. Plant Soil, 374: 251~262.

Sun L, Qiu F, Zhang X, et al. 2008. Endophytic bacterial diversity in rice (*Oryza sativa* L.) roots estimated by 16S rDNA sequence analysis. Microbial. Ecol., 55: 415~424.

Sung K, Corapcioglu M Y, Draw M C, et al. 2001. Plant contamination by organic pollutants in phytoremediation, J. Environ. Qual., 30: 2081~2090.

Swindell A L, Reid B J. 2006. Comparison of selected non-exhaustive extraction techniques to assess PAH availability in dissimilar soils. Chemosphere, 62:1126~1134.

Szmigielska A M, Van Rees K C J, Cieslinski G, et al. 1996. Low molecular weight dicarboxylic acids in rhizosphere soil of durum wheat. J. Agric. Food Chem., 44: 1036~1040.

Taghavi S, Barac T, Greenberg B, et al. 2005. Horizontal gene transfer to endogenous endophytic bacteria from poplar improved phytoremediation of toluene. Appl. Environ. Microbiol., 71: 8500~8505.

Tang J X, Liste H H, Alexander M. 2002. Chemical assays of availability to earthworms of polycyclic aromatic hydrocarbons in soil. Chemosphere, 48:35~42.

Tang J X, Petersen E J, Huang Q G, et al. 2007. Development of engineered natural organic sorbents for environmental applications: 3. reducing PAH mobility and bioavailability in contaminated soil and sediment systems. Environ. Sci. Technol., 41: 2901~2907.

Tang J, Wang R, Niu X, et al. 2010. Enhancement of soil petroleum remediation by using a combination of ryegrass (*Lolium perenne*) and different microorganisms. Soil Till. Res., 110: 87~93.

Tang Z Y, Wu L H, Luo Y M, et al. 2009. Size fractionation and characterization of nanocolloidal particles in soils. Environ. Geochem. Health, 31: 1~10.

Thygesen R S, Trapp S. 2002. Phytotoxicity of polycyclic aromatic hydrocarbons to willow trees. J. Soils Sedim., 2: 77~82.

Topp E, Scheunert I, Attar A, et al. F1986. actors affecting the uptake of ^{14}C-labelled organic chemicals by plants from soil. Ecotoxicol. Environ. Saf., 11: 219~228.

Torné J M, Claparols I, Marcé M, et al. 1994. Influence of pretreatments with inhibitors of putrescine synthesis on polyamine metabolism and differentiation processes of maize calluses. Plant Sci., 100: 15~22.

Totsche K U, Danzer J, Kögel-Knabner I. 1997. Dissolved organic matter-enhanced retention of polycyclic aromatic hydrocarbons in soil miscible displacement experiments. J. Environ. Qual., 26: 1090~1100.

Trapp S, Matthies M. 1995. Generic one-compartment model for uptake of organic chemicals by foliar vegetation. Environ. Sci. Technol.; 29: 2333~2338.

Trapp S, Matthies M. 1997. Modeling volatilization of PCDD/F from soil and uptake into vegetation. Environ. Sci. Technol., 31: 71~74.

Trapp S, Matthies M, Scheunert I, et al. 1990. Modeling the bioconcentration of organic chemicals in plants. Environ. Sci. Technol., 24: 1246~1252.

Trotta A, Falaschi P, Cornara L, et al. 2006. Arbuscular mycorrhizae increase the arsenic translocation factor in the As hyperaccumulating fern *Pteris vittata* L. Chemosphere, 65: 74~81.

Van Eerd L L, Hoagland R E, Zablotowicz R M, et al. 2003. Pesticide metabolism in plants and microorganisms. Weed Sci., 51: 472~495.

van Hees P A W, Jones D L, Jentschke G, et al. 2005. Organic acid concentrations in soil solution: effects of young coniferous trees and ectomycorrhizal fungi. Soil Biol. Biochem., 37: 771~776.

Vendan R T, Young J Y, Sun H L, et al. 2010. Diversity of endophytic bacteria in ginseng and their potential for plant growth promotion. J. Microbiol., 48: 559~565.

Viñas M, Sabaté J, José M, et al. 2005. Bacterial community dynamics and polycyclic aromatic hydrocarbon degradation during bioremediation of heavily creosote-contaminated soil. Appl Environ. Microbiol., 71: 7008~7018.

Voutsa D, Samara C. 1998. Dietary intake of trace elements and polycyclic aromatic hydrocarbons via vegetables grown in an industrial Greek area. Sci. Total. Environ., 218: 203~216.

Walter J, Weber W J J. 2002. Distributed reactivity model for sorption by soils and sediments: 15. High-concentration co-contaminant effects on phenanthrene sorption and desorption. Environ. Sci. Technol., 36: 3625~3634.

Wang M J, Jones K C. 1994. Uptake of chlorobenzenes by carrots from spiked and sewage-amended soil. Environ. Sci. Technol., 28: 1260~1267.

Wang Y J, Xiao M, Geng X L, et al. 2007. Horizontal transfer of genetic determinants for degradation of phenol between the bacteria living in plant and its rhizosphere. Appl. Microbiol. Biotechnol., 77: 733~739.

Weber W J J, Huang Q G. 2003. Inclusion of persistent organic pollutants in humification processes: direct chemical incorporation of phenanthrene via oxidative coupling. Environ. Sci. Technol., 37:

4221~4227.

Weber W J J, Kim S H, Johnson M D. 2002. Distributed reactivity model for sorption by soils and sediments: 15 High-concentration co-contaminant effects on phenanthrene sorption and desorption. Environ. Sci. Technol., 36: 3625~3634.

Wei Z G, Hong F S, Yin M, et al. 2005. Subcellular and molecular localization of rare earth elements and structural characterization of yttrium bound chlorophyll a in naturally grown fern *Dicranopteris dichotoma*. Microchem. J., 80: 1~8.

Weigel H J, Jager H J. 1980. Subcellular distribution and chemical form of cadmium in bean plants. Plant Physiol., 65: 480~482.

Welsch-Pausch K, McLachlan M S, Umlauf G. 1995. Determination of the principal pathways of polychlorinated dibenzo-p-dioxins and dibenzofurans to *Lolium-Multiflorum* (Welsh Ray Grass). Environ. Sci. Technol., 29: 1090~1098.

Wentworth C K. 1922. Scale of grade and class terms for clastic sediments. J. Geol., 30: 377.

West E R, Cother E J, Steel C C, et al. 2010. The characterization and diversity of bacterial endophytes of grapevine. Can. J. Microbiol., 56:209~216.

Weyens N, Vander L D, Artois T, et al. 2009. Bioaugmentation with engineered endophytic bacteria improves contaminant fate in phytoremediation. Environ. Sci. Technol., 43: 9413~9418.

White J C, Kelsey J W, Hatzinger P B, et al. 1997. Factors affecting sequestration and bioavailability of phenanthrene in soils. Environ. Toxicol. Chem., 16: 2040~2045.

White J C, Mattina M I, Lee W Y, et al. 2003. Role of organic acids in enhancing the desorption and uptake of weathered p,p'-DDE by *Cucurbita pepo*. Environ. Pollut., 124: 71~80.

White J C, Parrish Z D, Isleyen M, et al. 2005. Uptake of weathered p,pV-DDE by plant species effective at accumulating soil elements. Microchem. J., 81:148~155.

Wild E, Dent J, Thomas G O, et al. 2005. Direct observation of organic contaminant uptake, storage, and metabolism within plant roots. Environ. Sci. Technol., 39 : 3695~3702.

Wild E, Dent J, Thomas G O, et al. 2006. Visualizing the air-to-leaf transfer and within-leaf movement and distribution of phenanthrene: further studies utilizing two-photon excitation microscopy. Environ. Sci. Technol., 40: 907~916.

Wu N Y, Zhang S Z, Huang H L, et al. 2008. DDT uptake by arbuscular mycorrhizal alfalfa and depletion in soil as influenced by soil application of a non-ionic surfactant. Environ. Pollut., 151: 569~575.

Xie M J, Yan Z L, Ye Q. 2008. Effect of phenanthrene on the secretion of low molecule weight organic compounds by ryegrass root. Ecol. Environ., 17: 576~579 (In Chinese).

Xing B, Pignatello J J. 1997. Dual-mode sorption of low-polarity compounds in glassy poly (vinyl chloride) and soil organic matter. Environ. Sci. Technol., 31: 792~799.

Xu W H, Liu H, Ma Q F, et al., 2007. Root exudates, rhizosphere Zn fractions, and Zn accumulation of ryegrass at different soil Zn levels. Pedosphere, 17: 389~396.

Yang C H, Crowley D E. 2000. Rhizosphere microbial community structure in relation to root location and plant iron nutritional status. Appl. Environ. Microbiol., 66: 345~351.

Yang Y, Ling W T, Gao Y Z, et al., 2010. Plant uptake of polycyclic aromatic hydrocarbons (PAHs) and their impacts on root exudates. Acta Scicntiac Circumstantiae, 30: 593~599 (in Chinese).

Yang Y, Ratté D, Smets B F, et al., 2001. Mobilization of soil organic matter by complexing agents and implications for polycyclic aromatic hydrocarbon desorption. Chemosphere, 43: 1013~1021.

Yaws C L. 1999. Chemical properties handbook, New York. McGraw-Hill Book Co. 340~389.

Ying G G, Kookana R S, Mallavarpu M. 2005. Release behavior of triazine residues in stabilized contaminated soils. Environ. Pollut., 134: 71~77.

Yoshitomi K J, Shann J R. 2001. Corn (*Zea mays* L.) root exudates and their impact on ^{14}C-pyrene mineralization. Soil Biol. Biochem., 33: 1769~1776.

Yu F B. 2007. Isolation, characterization and bioaugmentation study of an efficient onitrobenzaudehyde degrading strain pseudomonas putida ONBA-17. Ph. D. dissertation, Nanjing Agricutural University, Jiangsu, China.

Yuan S Y, Chang J S, Yen J H, et al. 2001. Biodegradation of phenanthrene in river sediment. Chemosphere, 43: 273~278.

Zak D R, Holmes W E, White D C, et al. 2003. Plant diversity, soil microbial communities, and ecosystem function: are there any links? Ecology, 84: 2042~2050.

Zhang J Q, Dong Y H. 2008. Effect of low-molecular-weight organic acids on the adsorption of norfloxacin in typical variable charge soils of China. J Hazard Mater, 151: 833~839.

Zhang M K, Ke Z X. 2004. Copper and zinc enrichment in different size fractions of organic matter from polluted soils. Pedosphere, 14: 27~36.

Zhu L Z, Gao Y Z. 2004. Prediction of phenanthrene uptake by plants with a partition-limited model. Environ. Pollut., 131: 505~508.

Zhu X Z, Ni X, Liu J, et al. 2014. Application of endophytic bacteria to reduce persistent organic pollutant contamination in plants. Clean-Soil Air Water, 42: 306~310.

Zhu Y H, Zhang S Z, Zhu Y G, et al. 2007. Improved approaches for modeling the sorption of phenanthrene by a range of plant species. Environ. Sci. Technol., 41: 7818~7823.

Plates

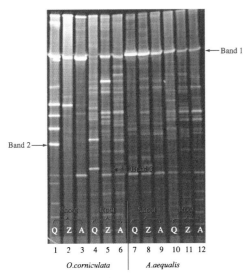

Figure 10-1 Representative DGGE for PCR-amplified 16S rDNA V3 fragments from endophytic bacteria in *Alopecurus aequalis* and *Oxalis corniculata*. The genus of the band marked in the figure: Band 1-*Pseudomonas* sp.; Band 2 - uncultured bacterium clone; Band 3- *Nesterenkonia* sp.

Figure 10-7 Fluorescence of strain Ph6-*gfp* in culture solution (×100). Bacterial cells with green fluorescence were displayed as bright green rods on a glass microscope slide

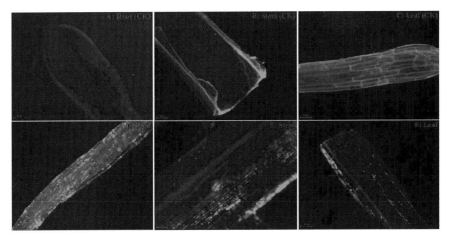

Figure 10-10 Visualization of inoculated endophyte Ph6-*gfp* within plant tissues. After inoculation with Ph6-*gfp* for 6 days, the colonization of Ph6-*gfp* was observed in plant roots, stems and leaves by fluorescence microscopy. Bacterial cells with green fluorescence were displayed as bright green rods inside roots (D), stems (E) and leaves (F). Plants with Ph6-*gfp*-free inoculation served as the control (A, B, C)

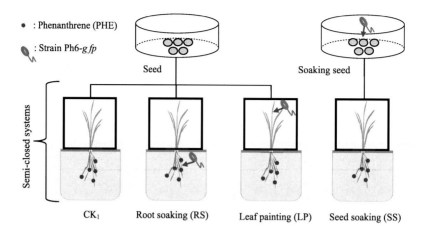

Figure 10-11　Diagram of the greenhouse experimental design in which a semi-closed system was used. Colonization methods were (1) seed soaking (SS), (2) root soaking (RS), and (3) leaf painting (LP). CK_1 indicates the control

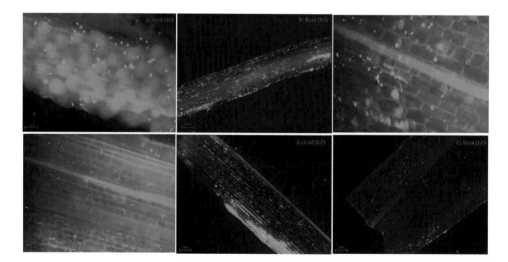

Figure 10-12　Endophytic *Pseudomonas* sp. Ph6-*gfp* cells in seeds, roots, stems, and leaves of inoculated plants were exposed to 1 mg/L of PHE via three colonization methods. Ph6-*gfp* colonization via SS was observed in plant seeds using fluorescence microscopy (A). After inoculation with Ph6-*gfp* via RS for 6 days, colonization was observed in plant roots, stems, and leaves (B~D). After inoculation with Ph6-*gfp* via LP, colonization was observed in plant leaves and stems (E, F)